For Tomorrow We Die

A Novel That Explores the Conflicts
Between Science and Religion

By
John Shelley

Revision 2.52

The Story:

Luke plans a family reunion with his estranged siblings in the Lakes Region of New Hampshire – and as he fears, conflicts over religious beliefs threaten to derail the gathering. Differences of opinion erupt into heated conversations as family members discuss the role of God, religion and science in their lives.

Topics they discuss include:

Life After Death
Heaven, Hell, Hades, and Purgatory
Evolution
Biblical Accuracy
Relativity
Salvation for Non-Christians
Ghosts and Spirits
Near-Death Experiences
Quantum Mechanics
Consciousness
The Rapture

The Family:

Matthew (The oldest brother): Matt is a physicist who is head of the science department of a small liberal arts college.

Deborah (Matt's wife): Deb is a psychologist that exhibits, at times, psychic tendencies.

Luke (The second brother): Luke narrates the story. Luke was attending medical school when a tragic event caused his life to spin out of control. He subsequently drops out of school, gets divorced, and loses all his faith and trust in God.

Naomi (Luke's second wife): Nan was raised in an orthodox Jewish home, but has had trouble balancing her strict Jewish upbringing with her own, more secular lifestyle.

Mark (The third brother): Mark was a prominent Methodist minister until the local paper published an article accusing him of having sexual relations with many of the women in his congregation. This eventually led to his disgraceful exit from the ministry, and Mark never regained his faith or trust in the church.

John (The fourth brother): John left home after high school because his father had no tolerance for his homosexual tendencies. He subsequently lost all contact with the rest of the family.

Mary (The youngest and only sister): Mary was a spoiled and rebellious child who left home as soon as she turned 18. She lived a marginal life over the years and was involved in many bad relationships – until she joined a fundamentalist church and became a born-again Christian

Contents

CHAPTER 1 – DAY 6

Try not to be a person of success, but rather a person of virtue.

Albert Einstein

The Mountain

I don't know if I can make it, but I have to try. My spirit is willing, but Lord knows, my flesh is weak. Just one step at a time – the right foot followed by the left – again and again – up the mountain.

I have spent my entire life trying to reach the summit. I have mounted countless expeditions, but I have had very little success. It seems every time I am gaining ground, every time my aspirations are within reach – I stumble, and fall back. And always, after each failure, after I muster the strength to pick myself up and try again – I inevitably fall once more. The mountain is stronger and more cunning than me. She tempts me. She entices me to scale slopes that are too difficult – way beyond my meager ability. And then, she strikes me down and beats me back – laughing at my feeble attempts to conquer her.

Why has life always been so hard? Why has fate always kicked me down and laughed in my face? Why is it that no matter how hard I try, I have never been able to become the man I know I can be? Why is it that my every attempt to climb higher has resulted in failure? Why do I always fall? Why is the mountain so steep?

It seems like my whole life has been wasted – just a disjointed accumulation of failed attempts, mistaken paths and lost opportunities. One disappointment after another. I have never been able to meet anyone's expectations – not my parents, or my wife, or my peers – not even my own.

I could have accomplished so much. I had so much potential, but it was swept away by an avalanche of circumstances and events that were beyond my control. I could have been a renowned surgeon, but after the accident I was forced to drop out of medical school. I could have had a successful business career, but my supervisors never understood my talents and I was unable to advance. I could have been a wonderful parent, but I was not allowed to raise the daughter that I loved so much.

Why so many obstacles? Why did I have to scale a slope steeper than anyone else? Why has the mountain always been against me?

The moon is almost full, lighting my path. I can see that the hardest part of the trail is ahead. Yet, I am undaunted because my goal is within reach. Just a little further, and I will be at the top of Rattlesnake Mountain.

I have climbed this mountain before – when I was younger, but I have never seen a sunrise on the summit, and this might be my last opportunity. I must continue. It may just be a small victory, but I can do this. No more failures. There will be strength and solace in the morning light once I reach the top, and nothing is going to keep me from that. This beaten and forsaken body must move on. And so, I take a few deep breaths, and forge onward with an inner determination that transcends pain and fatigue.

I grab a discarded walking stick from the side of the trail to help me up the path, and I climb higher and higher by anchoring my feet on exposed tree roots crossing the washed out trail, and on granite stones that were randomly left behind from the last ice age. My cramped legs are on fire, and my breathing is strained and erratic, but I am gaining ground.

"Just one step at a time."

I have to stop every five or ten minutes for a short rest, but then I continue forward.

"I can do this," I repeat over and over to myself, and my confidence is building as I finally near the top of the trail. But then, once again, like so many times before, fate strikes another blow.

Suddenly, my left foot slips on a moss covered rock and my ankle twists in the soft clay on the side of the trail. I fall back, grabbing a small sapling to try to regain my balance. For a split second, I think I have averted a disastrous fall, as I cling precariously to the small tree, holding my breath as a single bead of sweat drips off my brow.

But the mountain is conspiring against me. She knew I was coming, and had carefully placed the slippery rock and the wet clay along the way, and now, she has given me a young tree much too weak to support my weight. The sapling breaks, and I careen backwards, head first into an immovable century old oak. The majestic tree holds me up for a moment, but then gives me back to the mountain. I collapse to the ground, semi-conscious, and roll down the slope, finally being stopped by an old, soft, rotted log crossing the path.

As I lie there on the damp, dew soaked ground, I am overcome by sadness and a sense of resignation, for I realize that my climb has ended. Another failure. Another failure heaped on top of a lifetime of countless other failures.

But now, perhaps, this is the end. My final failure.

Rattlesnake Mountain is not a difficult summit to scale. Many kids in their late teens and early 20's routinely jog up most of the trail, sporting their slim torsos and toned muscles. Parents in their 30's sometimes carry their sobbing and overstimulated children to the top in backpacks. Old guys, hoping to delay their inevitable decline, can often be seen hiking along the way, resting periodically on old tree stumps. It's considered a relatively easy, family friendly climb.

And there are no rattlesnakes in this part of New Hampshire. The mountain received its name from the bare rocks at the top, which, if you have enough to drink, or a good imagination, looks a little like a rattlesnake tail.

The mountain is a popular place for locals and tourists alike, mainly because the view from the top is spectacular. From on high, you can see the entire expanse of Squam Lake, and a panorama of hills and islands that is almost dreamlike. Such a beautiful place. You can feel close to nature – and close to God. Something I so desperately want to experience.

But it isn't to be. I cannot move. My head is throbbing and my ankle is swelling. The mountain has dealt a final blow to an already broken spirit and broken body.

As I lie there, trying in vain to collect my thoughts and evaluate my condition, my mind starts to wonder, and I begin to visualize my life as it could have been. I see myself at the seashore with my wife and children. Lots of children. The children I have always dreamed of having. We are all running through the ocean waves with abandon, with no cares or worries. A wonderful time. I see us building sand castles together – fortresses to withstand and rebuke the rising tide, and to guard and protect this family in times of crisis and uncertainty.

And then, I see us all together as we gather arm in arm around an evening bonfire, and we feel the glow of the flames warming our hearts, and forging a strong, silent, inseparable bond between us. And we all feel united by a binding love – a strong and sustaining love that triumphs over adversity and fear.

It's the way it should be. It's the way it was meant to be. It's the family I have always wanted.

But it is too late for that.

And a chill runs through the entire length of my body, as my dreams fade into reality and my delusions of a better life drift away. And I am faced with the realization that there is precious little time left – far too little time to change course – to make something better of my life. Too little time to correct past failures – no more opportunities to climb higher.

The battle is almost lost, and I am much too weak to continue the fight. The odds seem insurmountable and there are no reinforcements in sight.

The enemy is advancing. I sense that the cancer that I have been fighting for so long is slowly invading every major organ in my body. It's on the offensive and marching relentlessly forward. There is no stopping it, and as I close my eyes, I begin to visualize a whole battalion of new tumors growing in my liver, and in my lungs, and in my brain – gaining strength, minute by minute – slowly and deliberately taking control. I can hear my tissues groaning from the assault, and I can feel my muscles weakening as the invading cancer saps my own blood supply for its nourishment.

Those crafty little lymphoma cells – circulating, probing – looking for new locations to invade and destroy. Little terrorists that have been able to evade the coalition forces of chemotherapy, steroids, and monoclonal antibodies. The resilient bastards have been in remission for 7 years, but they have survived by hiding in clandestine caves and crevices deep inside my bones. And now, they are mounting a new offensive, with a mutated arsenal of weapons that are no match for this battle weary soul.

Maybe it's just time to surrender, and admit defeat. It doesn't make sense to fight a losing battle – one with no hope of success. What does that accomplish? It's time to face reality.

My days of climbing are over. The mountain has won the final battle. I finally realize that I will never be able to live my dreams – to experience the life I have always strived for. The climb has just been too difficult.

Even my hopes of seeing a sunrise on the summit of this beautiful mountain have been thwarted. Perhaps a small setback, but one indicative of so many others. Just one of the many hopes and aspirations that have eluded my grasp.

My life has been a failure ... and I realize that time has run out.

I have fallen for the last time. And so, I rest my head on the soft, leaf covered ground and close my eyes.

I am ready to die.

The Stranger

"Com'on Luke. Let's get up. It's not your time yet."

I open my eyes. I am dazed for a moment, not able to remember where I am. But as my blurred vision begins to sharpen, and the aching in my head and ankle become more acute, I remember my failed climb, and I realize I am lying across the trail, blocking the path for everyone else.

I see in front of me a short portly gentleman, with a white beard, wearing overalls and a lumberjack shirt. It seems lighter now, as I can clearly see his weather worn face and his benevolent expression. It is the type of face that is hard to forget – strong rustic features, but with pudgy cheeks and kind eyes.

"How long have I been here?" I wondered.

"Looks like you had a pretty bad fall Luke, but I think you're gonna be OK."

"Who is this guy," I think to myself, "and how does he know my name?"

"It's not that much farther to the top. I think you can make it. Let's go."

The kindly gentleman helps me up, and as I reach my feet, to my amazement, I feel better. Maybe the fall was not that bad after all. Strength has returned to my legs and my lungs feel refreshed in the cool morning air. I manage to speak for the first time, "You from the camp?"

"Oh no," says the gentleman, "but I've lived in these here parts for a long, long time. Ready?"

He puts my arm on his and we begin walking the same path that had defeated me earlier. Seems easier now though. It's amazing what a little rest can do to rejuvenate the body. We walk past the rock and the clay and the sapling, defying the mountain's previous advances against me – up the steepest part of the trail with minimal trouble. And as we continue our climb, I begin humming a marching tune that I remember from my youth. I don't remember the song's name or the words, but I think it was a piece written by John Philip Sousa in the late 19th century. And as I begin to hum louder, I lift my legs higher and begin stepping in time with the slightly off-key music. I'm sure I look a little silly, trouncing up the trail like an adolescent drum major in a high school band, but my new friend doesn't seem to mind. He just keeps marching in step at my side, and he looks over, and smiles at me, with those soft eyes and rosy cheeks, and he nods, as if to say that he is enjoying my company – and, perhaps, my singing. And I begin to feel a sense of joy, and peace and accomplishment.

And then, after sidestepping some loose rocks, we round the final bend on the trail, and I can see the clearing at the top of the mountain. And as we pass the last trees, I turn with eager anticipation to view the majestic lake below.

"Seems darker now," I think to myself. "I can't really see much." I then look to the horizon and notice just a faint glimmer of light in the east. The sun is just beginning to rise. "I made it! I made it!" I mumble to myself out loud. Finally, for the first time, I'll be able to see the sun rise from this precipice – to witness the majesty of God's creation as soft rays of light slowly illuminate His handiwork. It's a heavenly vantage point, far above the cares and worries of the lower realms, and I stand motionless for a moment, as if in a trance, as I stare out toward the horizon.

"I can't believe I made it! I can't believe I made it!" I keep repeating over and over. Success at last. For once, the mountain has been defeated. I know it's just a small accomplishment, but I savor the victory, and I anxiously await for the booty of battle – to be able to see a new day dawning, at one of the most beautiful places I have ever known.

And maybe, just maybe, this new day marks a new beginning. The day I finally begin to turn things around. The day I begin to put my disjointed life back together. The day I seize the opportunity to leave my past failures behind, and march forward, arm in arm with the people in this life that matter most to me.

I turn to the benevolent stranger who helped me up the hill – to thank him for his help – but he is gone.

"There will be time to thank my new friend later," I think. "He won't be hard to find. He's a local who has lived in this area for a long time, and all the locals know each other. I'll talk to Joe at the General Store in Holderness. If he doesn't know him, I'm sure he knows someone who does.

"Funny," I think, as my mind begins to wonder, "It seemed so much lighter when the kindly gentleman picked me up off the ground. I was sure I had missed the sunrise. Maybe it just appeared lighter because he had a lantern. I don't really remember. I know I grabbed hold of his left arm. I think it was his left arm ... or maybe it was his right. Anyway, I don't think I saw what was in his other hand. It could have been some type of light. I was still a bit dazed as we walked up the mountain. I don't really remember much about our climb together, or what we said to each other – or if we even talked at all.

"And why was it so much darker when we reached the summit? Perhaps my friend turned off his lantern after the final turn on the trail. I think we came around that final turn together. Yeah, I'm almost sure we did. We must have. He must have turned off the lantern after that turn. That is, if he had a lantern. Or, maybe it was darker because the moon had set, or had gone behind a cloud. I think the moon was out. Yeah, I know it was.

"Or, perhaps," I think as I try to concentrate and focus my eyes on the horizon, "maybe that bump on my head has impaired my senses and affected my memory. My ankle feels better but my head is still throbbing. I wonder if I have a concussion. Maybe my eyes are dilated. I probably should see a doctor as soon as possible."

But then, my thoughts return to the lake, as the sunrise receives my full attention, and hews of red, orange and yellow begin to emerge in the eastern sky.

As the light becomes stronger, I begin to see the clear calm waters of the lake, reflecting the sky above and the trees below. So peaceful. So surreal. Islands begin to emerge from the mist as the sun peaks out from behind Red Hill. It's a cinematic wonder with swallows, warblers and ducks providing the background music. It's as beautiful as I could have imagined.

"If there is a God, he must be here somewhere. If only I knew where."

I subconsciously find myself repeating a verse I learned as a child in Sunday school.

He leadeth me beside the still waters.

It's from some Psalm. I don't remember exactly which one, but I know they always recite it at funerals – and I have been thinking a lot about funerals lately.

The rising sun warms my body, as I soak in its energy and strength.

He restoreth my soul.

A light breeze begins to blow across the lake causing tiny waves on the water. The sun's rays landing on these ripples create an array of millions of small star-like reflections across the pond. The dazzling display makes we wonder, "Why is the light so comforting? Is it just its radiance, or perhaps the warmth it brings – or is there more? They say God is light. I know that's just a metaphor, but maybe there's more to it."

This week has given me a new appreciation for the power of light, or maybe the power in the light.

I close my eyes and turn my face toward the sun, trying to feel the energy – the life in the light. But after a few minutes, I turn away, unable to sense anything out of the ordinary. I have learned so much, but there is still so much that I don't understand.

There is, however, one thing I do know. I no longer have an overwhelming fear of death. Dying doesn't scare me as much as it once did.

Yea, though I walk through the valley of the shadow of death, I will fear no evil.

"It's funny," I think to myself, "that no one ever thinks about death, or fears death – until they are actually facing death. And then, from that time on, it consumes your whole life. All your thoughts, all your activities, all your conversations – all your nightly dreams."

I begin thinking about my funeral. "What are people going to say?"

"He was an OK guy, even though he never really put it all together." "He provided for his family, as best he could." "He loved his dog."

"Maybe I should write a script for everyone to follow. That will keep the funeral from becoming too embarrassing. They can read that psalm, say a couple of exaggerated pleasant things, have some fat lady sing Amazing Grace – and then leave. Short and sweet is probably best.

"Why does it matter anyway? I won't be there. I'll be in ... well, somewhere. Maybe, I guess. I'm not really sure anymore."

... and I will dwell in the house of the Lord forever.

There are so many things going through my mind. Up until this week I had always thought that once you died, it was over. Just a decaying body to be buried or burned. And that's it. Dust to dust. No immortality. No eternal soul. Done.

Now, however, I'm confused. I'm a practical guy, and my training in the natural sciences and my medical background had taught me to believe that science could explain all things, given time. But now I don't know what I believe.

I close my eyes for a moment, and try to construct a prayer asking God, or someone, or anyone – for understanding, or help, or comfort. But I become frustrated with the attempt. I can't find the right words. "This is useless," I think to myself. "I don't know what to ask for. I don't even know who I am trying to talk to."

As I open my eyes, my attention returns to the lake, and as I peer down, I notice how different things appear when you look at them from on high. There is a completely different view of the world from above.

I keep hearing Matt say, "Things aren't always as they appear," over and over again.

Off in the distance, I can see Birch Island where we searched for wild blueberries, and I can see that there is a large grove of blueberry bushes on the south side of the island – bushes that we overlooked earlier this week while scavenging for enough fruit to make a pie.

And, further out, I see Church Island where we had the wedding, and I notice how the path winds from the boat dock on the south shore to the outdoor chapel on the east, where I can see the chapel's cross, made from white birch tree limbs, rising tall and strong above the rocks on the point.

To the east of Church Island I see a canoe in the water with two teenage boys. They are feverishly paddling toward Little Loon Island to catch a glimpse of the bald eagle nest in one of the trees on the small atoll, but they are so focused on reaching their destination that they miss seeing the magnificent raptor swoop down behind them to snatch an unsuspecting fish from the water. If only they could see what I am seeing from here.

Then, I hear the sound of a motorboat breaking the morning silence, and I turn to see a fisherman, steering his Boston Whaler directly toward the shallow rocks near Otter Reef. I can see he is in imminent danger of running aground on the submerged boulders surrounding the reef, and I instinctively yell, "Careful!" even though I know my warning will be in vain from this distance. But the boater's view is different than mine. He sees an open lake, and a smooth ride to a prime fishing hole. But I see that he is in danger. There is, however, nothing that I can do, and I watch as his outboard motor strikes a shallow rock. I am concerned for a moment, but the fisherman appears to be OK. He restarts his stalled motor and slowly returns to shore, his boat vibrating a bit because of a damaged, unbalanced propeller.

It's a different view of the world from up here. This week I have come to appreciate that the things we see, the things we experience, the things that we are sure of – may not be an accurate description of reality. I am going to have to start looking at things from a different vantage point – from higher up.

I keep hearing Matt say, "Things aren't always as they appear."

And, as I continue to stare down at the lake below, I begin to wonder what things are like from God's vantage point – how different His view of this world is from mine. Could it be that He sees my obstacles, my troubles and my afflictions in a different light – from a different perspective? Maybe He's not the aloof absentee God I imagine. Maybe he just has a different point of view – from higher up.

"But that's just speculation," I think to myself. "I'm still looking for God."

I begin to smell the faint scent of burning wood, and I look down to our camp directly below. Smoke has started rising from a few of the cabin chimneys, as the camp begins to awaken to a new day.

It's called Deephaven Camp, and it's where my three brothers, my sister and I have spent the last week – a wonderful place, with so many fond memories. I can see the ball field where we played capture-the-flag as kids, and the tennis courts where we first learned to swing a racquet, and the woods, where we roamed through patches of poison ivy chasing our errant tennis balls. Then, looking a little harder, I see the dining hall hiding beneath the tall pines, where we would eat ice cream before and after lunch – and dinner. And over on the point, on top of the rocks, a few feet from the water, the cabin where we would always stay. I remember waking up on cool, clear mornings just like this, to the sounds of chirping birds and a roaring fire, looking out over the clear spring fed waters of Squam Lake. It's a hypnotizing scene that can even captivate the spirit of young boy, who, at the time, cared more about baseball than nature. There's no place like it. The combination of natural beauty and childhood memories is rejuvenating – a little bit of heaven here on earth.

Our parents brought my three brothers and me here five or six times when we were kids, but that was over 40 years ago – before Mary was even born. Funny though, not much has changed. Still no phones or TVs in the cabins. No heating or air conditioning. It's a place where you are forced to lead a simpler life. A place where time inevitably slows down, and affords you the opportunity to notice and appreciate things you often overlook in a faster paced environment – important things – like nature and friends and family. It's a place where worries can be temporarily forgotten, and where relationships can be restored and nurtured. It's a place I have longed to come back to – for a long time.

The first campers came to Deephaven in 1897, a year after its founder, Alice Mabel Bacon, purchased the property for $500. That year, the campers themselves helped build a dining room, a kitchen with a tent roof, and the first cottage. The camp grew slowly, but steadily throughout the years, with many families building their own cabins during their summer-long stays.

Most early camp visitors would arrive by train in the nearby town of Ashland. They would then pile their luggage on a horse drawn wagon for a short ride to a dock on Little Squam Lake, where they could catch a small steamboat to Deephaven. When better roads were built and cars became more commonplace, Miss Bacon insisted that all vehicles driven to camp be parked at the Deephaven entrance, and not be allowed to travel any further because they made too much noise, and would distract from the natural beauty of the lake.

There has always been a delicate balance here between simplicity and progress. The camp has always strived to maintain a natural setting where extended families can return year after year to rekindle bonds weakened by work, and routine, and technology.

But progress is inevitable, and simplicity is often inconvenient, and so, eventually over time, modest improvements were made to the camp. The cabin outhouses were replaced with indoor plumbing, and kerosene lamps gave way to electric lights. The cottages have had rooms added, porches extended and foundations reinforced, but many of the original cottages remain today, as evidenced by the squeaking wood floors and drafty windows.

We have been here since last Saturday, almost a week. A family reunion of sorts. I haven't seen my three brothers, Matt, Mark and John, or my little sister Mary for quite some time, and this was a convenient time for us to get together. My wife's daughter, from her first marriage, was married on a small island in the middle of the lake on Wednesday, and my brothers thought that it would be fun to spend a few days together, both before and after the wedding, in a place we loved so much as kids.

Never, in my wildest dreams, did I ever expect it to turn out like this.

It all began last Saturday.

CHAPTER 2 – DAY 1

The value of a man resides in what he gives and not in what he is capable of receiving.

Albert Einstein

Logan Airport

"Just take a couple deep breaths and you'll be fine," says Nan robotically, as she continues to read her book. "We'll be on the ground in a few minutes." It has become routine for her, the same words – every time. And just as every time before, it is of little comfort. My fingernails grip the armrests, leaving indentations in the leather. For me, it's always the worst part of the flight. The landing.

I can't help it. When I hear the words, "Please return your seat backs and tray tables to their normal and upright positions," my palms begin to sweat. Then my stomach starts to feel queasy as the plane begins to decelerate and lose altitude. My legs become tense and my shoulders freeze as we rattle through the clouds, and accelerate and decelerate and accelerate and decelerate, and then, as the flaps and the landing gear are lowered, and the cacophony of noise from the increased air resistance fills the cabin, and the plane descends perilously close to the ground – I grip the armrests, and look straight ahead.

"Are we there yet?" I ask, just like a child.

"Just a minute more. You really need to read this book. It's great!"

The plane approaches the airport faster than normal because of strong crosswinds, and hits the runway a little harder than expected. There's a small jolt as the wheels touch, and few passengers gasp, but I barely notice. Just the thought of being on “terra firma” is comforting to me, and my muscles begin to relax.

“You really need to do something about your fear of flying, Luke.”

“What do you suggest Nan?” I reply a bit sarcastically.

“You could drink more.”

“Yeah, but it's only 10:30 in the morning – and I have to drive.”

“Good point. Maybe I should drink more.”

After the plane arrives at the gate, I retrieve our carry-ons from the overhead. And then we wait – standing impatiently in the aisle as the girls soccer team in front of us collect their belongings from on the floor, and from the seat back pockets, and from the overhead bins, and from friends across the aisle. It seems like ten minutes has gone by when I blurt out, “Can you try to hurry? We have to meet people from another flight.”

Matt, my older brother and his wife, Deb, are scheduled to arrive within a few minutes of our flight, and our plan is to meet each other at Boston's Logan Airport, rent a car, and drive to New Hampshire.

Nan just looks at me, slightly raises one side of her mouth in a half smile and shakes her head. I know what she is saying without any verbalization. I know I need to practice patience. I know that there is flexibility in our schedule and I know we have plenty of time – but I am still a bit anxious and nervous from the flight. I just want to get off the plane.

The aisle finally clears and we walk through the front cabin door, down the jet way and into the terminal. Nan and I then hurry to the nearest flight monitors. “What's their flight number?” inquires Nan.

"I don't know. It's coming in from Philadelphia. Right about now."

"Here it is. It's ... ah ... delayed. About an hour and a half. Good thing we hurried."

"OK, OK." I say, trying to dispel my unrelenting urgency. "I guess we can pick up the rental car and then come back to the terminal and wait for their flight to arrive."

We stop for a minute at a coffee shop in the terminal where I, as I always do, order a large black coffee. Nan, as she always does, orders something I have never heard of before. "What is that?" I ask despairingly.

Nan just pats me on the back and says, "It doesn't matter Luke. You don't have to drink it."

Nan and I have been married for about 7 years now, and I am still amazed, daily, at how well she tolerates me. All my little idiosyncrasies, all my phobias, all by babbling. They don't seem to bother her. She just accepts me for who I am, and goes on about her life. Must be love. I can't explain it any other way.

We met at the hospital, during the time I was battling cancer, and having trouble dealing with the seemingly endless rounds of chemotherapy. She was the one that pulled me through. She was the one that gave me hope when there was no one else to offer support.

When Nan found me, I was just a shell of a man – almost empty, filled only with fears, anxieties and self-pity. I was so afraid I was going to die. I couldn't work because I was often too weak to get out of bed. I couldn't sleep because of the recurrent nightmares, and I couldn't eat because the chemo had ravaged by stomach. Essentially, I couldn't do anything but worry about my disease – and that's what I did. All day long.

Nan worked as a nurse for my oncologist, and she took me on as her personal project. She had seen so many just like me before, and she knew what was required for my rehabilitation. She refused to allow me to feel sorry for myself. She told me that I was "blessed" to have a disease which was curable, and she told me countless stories about patients who were less fortunate than me.

And then she said something I will never forget. She said, "Luke, your main problem is not that you may be dying – your main problem is that you aren't living. The reason you are so afraid to die is because your life, up to this point, has been an unfulfilling, miserable existence – and dying would be an admission that you had failed at life – an admission that you had not lived your life to the fullest, the way it was meant to be lived."

I had known for some time that there was something wrong with my life, but I was never able to pinpoint the exact problem. I tried to blame my troubles on others – my first wife, my boss, my stock broker. I was so blind. That is, until Nan so poignantly identified my real problem. And at that very moment knew she was right. It was as if she had hit me over the head with a 2 by 4 and knocked some sense into me.

Nan told me that I lacked something called *gemilut chasidim*. It's an old Jewish term which means bestowing loving kindness or acts of compassion. It's an obligation to God, a mitzvah to help others. She said that the main thing missing from my life was a genuine concern for other people. She told me that the only way I would be able to fully recover, the only way I would be able to beat the monster raging inside me, was to begin to care less about myself and to care more about others. To her, the chemotherapy was just a secondary treatment. What I really needed to be cured was the realization that I was part of a community that needed my participation, my help and my support. She told me that I needed to build a bridge from the island I had formed around myself, and walk across.

I knew she was right. The reason I felt so helpless fighting this disease, the reason I was so afraid ... was because I was all alone. There was no one to lean on in my time of need, because I had not taken the time to nurture any long term relationships – at least not for many years – since the accident. I have always been so self-sufficient. I didn't need anyone else. I learned to cope and adapt to every tragedy in my own life, and I did it by myself – without help from anyone else. And since I was able to manage on my own, through unusually difficult circumstances, I always thought that others ought to be able to do so as well. But look where it got me. Nan said it best, "a miserable, unfulfilling existence."

Nan's solution to my problems was simple, just a few small changes in attitude and behavior. But, as it turns out, they were changes that have been very difficult for me. I have always thought about myself first, and it is so hard to reverse a lifetime of self-indulgence. Progress has been slow. I have tried to think more of others, but it has been a tedious learning process, even after the endless hours of volunteer work and community service. I still find myself reverting to my former self. I can't help it.

Part of my problem is simply that I am so impatient, and learning tolerance has been difficult. It's an attribute I was not born with, and developing a skill that is foreign to my DNA has had its setbacks.

Another obstacle I have not been able to overcome, is my fear of dying. It still haunts me daily.

But I have made some progress. I did learn to genuinely care for someone – a close friend – in an unselfish giving relationship. It was something I had not been able to do for a long time, and it felt so good. The funny thing is, I didn't see it coming. It just happened. I fell in love.

Nan and I were married not long after I finished my treatments – before my hair had fully re-grown. Since it was a second marriage for both of us, we planned a small wedding – just a few mutual friends and a couple relatives. Only a few members of Nan's family attended because they didn't fully approve of our relationship. Not that they had anything against me personally, but Nan is Jewish, and I am not, and an interfaith marriage was not acceptable to many of her relatives.

But it really didn't matter. No one from my family attended the wedding at all. My mother and father had died some time ago, and I had lost track of most our aunts and uncles. I invited my three brothers and my sister, but they all lived so far away. And besides, we were never very close as children. We were all so different, and time and geography had further separated the already weak bonds between us.

We take the escalator down to baggage claim and start waiting again. I begin wondering how we are going to be able to wheel all the luggage out to the rental car bus. I only brought one medium sized suitcase, but Nan brought three large bags. Of course, that did not surprise me. Nan has always had trouble packing light, and I knew she was going to bring three or four dresses for her daughter's wedding, because she was having trouble deciding which one to wear. Fortunately, I had planned ahead and reserved a full size SUV at the rental car agency. The largest one available.

After a few minutes of waiting, Nan pulls out her book and begins reading – and I begin pacing. "How long can it possibly take to unload a few bags? This is ridiculous," I say loud enough for everyone around to hear.

“Remember Luke, you are on vacation.”

“Yeah, but the airline doesn't know that. They should be more responsive.”

“Why don't you complain?”

“I am complaining!”

I don't know why I am always so impatient. I have been working so hard at being better. Nan says that I have “hurry-itis” – a disease where you constantly rush through life, never taking time to enjoy the journey. She's right, as usual, but there is just something in my nature that admires expediency and efficiency – and it's hard to enjoy a journey when that jerk in front of you is driving so slow.

About that time, the conveyor honks and bags begin to appear from luggage purgatory somewhere below. My suitcase is the fifth or sixth one out, and I pull it off the belt and efficiently attach my carry-on. I'm ready to go. But Nan's three suitcases take a bit longer. Finally, I recognize one of her overstuffed bags lumbering toward me and I wrestle it off the belt with my left hand. Nan grabs the second one while I am still rustling with the first. Looking around, I notice her last bag on the other side of the carousel. It was so large it had been unable to navigate a turn in the belt and had become stuck on a small rail while rounding the bend. I run over to rescue the beast, as bags are beginning to pile up from behind – but in my haste, as I grab the handle and begin to pull, I spill my coffee all over the conveyor belt, soaking two adjacent suitcases.

“Not again,” I think to myself. “How is it that I always manage to do something awkward when Nan is watching?”

Nan just shakes her head, and helps me apologize to the two New Yorkers who were unexpectedly, very understanding about their wet luggage.

“Hey, dat's nothin', just a little bit of coffee”, says one lady.

"You should see how bad they treat bags at LaGuardia," quipped her friend.

We then corral all the suitcases together in a circle and formulate a plan to exit the terminal as quickly as possible, to avoid any further embarrassment. Nan agrees to take my bag and carry-on, and I agree to take the rest. I throw my empty coffee cup towards the trash can, but miss by a few inches, and have to pick it up. Nan is already ahead of me, but I am delayed a little longer, trying to gather momentum as I push one of her bags in front of me and pull two from behind, with Nan's carry-on on my shoulder. I feel a little unbalanced as I make a beeline for the door, swerving to the right and left, trying not to trip over anyone on the way.

I am out of breath and a little dizzy when we reach the bus stop, and I fall on top of one of the suitcases as I try to wipe the sweat off my brow with my shirt sleeve. Nan offers her hand to help me up. "Sorry," she says, "I could have carried more."

"That's OK."

Fortunately, the bus arrives within a few minutes, and the driver is very helpful, lifting all our luggage onto the bus, as I take a seat in the back and try to rest for a minute. "I just don't have the energy or the stamina I used to." I think to myself. "I guess I'm just getting old."

We arrive a few minutes later at the car agency and drag the suitcases one by one to the sidewalk. "We're almost there", I say under my breath.

But today, nothing is going to be easy. My name is not displayed on the priority customer marquee, and I know that that usually means there is a problem. "I can't believe they don't have our car ready." So, we ditch the bags on the sidewalk and go inside to the rental counter ... and wait. The line is 3 people deep and there is only one agent.

Nan knows that I am not very good at waiting. It doesn't matter where – at the bank, at DMV, on an airplane, at baggage claim or at a car rental office. I start tapping my foot and looking repeatedly at my watch. "I can't believe they put their customers through this." I then turn to check the suitcases outside. They are still there, but there is some guy out there looking around. He looks suspicious. "Nan do you think you should go out and watch the suitcases?"

"I would Luke, but I'm reading this book. You really need to relax."

A short while later, as I am looking back and forth between my watch and the suitcases, I hear a pleasant young voice say, "Good morning sir. May I help you?"

It was my turn. The wait wasn't as long as I had thought. So, I gather my thoughts and as I am marching up to the counter and about to say, "You guys screwed up my reservation, again!", I think that maybe I should try a different approach – the way Nan has tried to teach me. So, mustering all the social grace that I have in reserve, I say, "Can you check on our reservation, please."

The agent smiles, and says, "May I have your confirmation number."

I pull out the reservation form from the side pocket of my carry-on and rattle off the numbers. Then I pick up a nearby pen and begin tapping on the counter as we both wait for her computer to pull up my information.

"Oh ... let's see ... here it is. I have a reservation for a full size SUV for one week." she finally replies.

"That's it."

"But the reservation is for September, a few months from now. This is June."

The comment infuriates me, and as I am about to make a disparaging remark about how I know what month it is, and as I am about to shove my confirmation in front of her face and tell her to find the correct reservation in her stupid computer, and as I am about to ask to speak to the manager ... I notice that the pickup date on my printout is ... in

September. It was my mistake. I had made the reservation online, and apparently, I hit a wrong key. I entered a 9 instead of a 6 in the month field – September instead of June. "I can't believe I did that."

Nan just shakes her head.

"Do you have any big SUVs available?" I ask sheepishly.

"Doesn't look like it. Not for the whole week."

"How about a small SUV?"

"Nope, they're all gone. I can give you a luxury car for the same price."

"How many bags can that hold?"

"Oh, three or four, depending on how big they are."

"That's not going to work. We have 4 people and probably 6 suitcases – plus carry-ons."

"Well, how about a minivan?" There's plenty of room for 4 people, and you can lay the back seat down and pile all the luggage to the ceiling. Large families do it all the time. It works really well."

"I don't want a minivan."

"I'm not sure you have a choice. It's the only thing we have available that will meet your needs."

"OK. OK. Fine. It's not blue, is it?"

"I don't know, but this is the last one available. Want it?"

"Yeah."

We complete the paperwork and get the keys. Fortunately, the car is close – just a few spaces from where we dropped off the suitcases. It is a white van which appears to have been abused by its former renters. There are numerous scratches in the side and a fairly large dent in the front bumper. As I pile all the luggage into the back, Nan fills out the scratch and dent form that was included with our rental contract, and we drop it off to the attendant as we exit the garage.

As we head back to the airport, I fumble with the radio until I find an oldies station playing 60's and 70's music – my favorite era. The first song to play is "Jumping Jack Flash" by the Rolling Stones, and I, as I usually do, began singing along. Nan knows how badly I sing, but over the years she has learned to tolerate the discord. Sometimes she sings along, but usually she just watches the road – to help me avert accidents when I become distracted by the music.

"I'm Jumpin' Jack Flash. It's a gas! gas! gas!" I bellow.

Sometimes, on long trips, Nan brings along earplugs.

We arrive at the terminal, park and walk over toward the baggage claim area. As we are entering the front doors, we are passed by a family facing a struggle similar to the one I had had earlier. The father is trying to cope with three suitcases, a portable baby crib and two car seats, while the mother is holding an infant, two carry-ons and a diaper bag, with a toddler grasping to the hem of her dress. Both parents are unaware that the toddler had dropped a small blanket she was carrying and they continue walking toward the rental car bus lanes.

Nan elbows me in the side and says, "Why don't you see if you could help them out. It would be a nice thing to do."

I pause for a moment, preferring not to become involved with a couple of complete strangers in the middle of a major metropolitan airport, but at Nan's prodding I pick up the blanket and walk outside and return it to the little girl.

The mother and father are both appreciative. "She probably would not be able to sleep tonight if she had lost her blankie," the mom says. "Thanks."

I then help the father load the luggage, the car seats and the crib onto the bus.

When I return to the terminal I find Nan sitting in a chair, watching me through the window. I sit down next to her and say, “You spying on me?”

Nan reaches over and gives me a small kiss. “Yeah. I am. You're doing better, but it's been seven years now, and you know, you still have difficulty thinking about the needs of others. It's not spontaneous for you. You have to be pushed before you offer assistance, or become involved in social situations.”

“I know. I am trying.”

“Yeah, I know you are. But you need to learn to take the initiative – to jump in and offer help whenever you see the need, and to get involved with others, even though you may feel a little uncomfortable.”

“I know. I know. I'm working on it. I really am.”

Matt and Deb

We had spent so much time at the rental car agency that Matt and Deb's plane had already landed by the time we return to the terminal. As we wait for them to arrive at baggage claim, I pull out Matt and Deb's last year's Christmas card from my carry-on. It has a picture of them and their dogs in front of their home in New Jersey. Nan has never met my brother or my sister in-law, and it has been a good 10 or 15 years since I have seen either of them.

"This is what they look like," I say, showing Nan the photo.

"Funny," replies Nan, "they look so much alike."

It has never really occurred to me before, but Nan is right. They look so similar. They both have graying shoulder length hair, and they are both about the same height and have slightly rounded stomachs. But there is more. Their eyes and their smiles are almost identical, and those are things you can't change with a haircut or a diet.

"You know, I don't ever remember them looking like each other, but I haven't seen them in a long time."

"Maybe, it's just the way the photo was taken," replies Nan.

"Oh, there they are."

I wave to Matt and Deb as they are descending the escalator, and I put their picture back into my carry-on as they walk toward us, side by side. "It's uncanny," I think to myself, "Nan was right. There is an amazing resemblance."

I shake Matt's hand and then follow with an awkward half hug. "Let me show you how this is done," says Deb, as she wraps her arms around me and pounds me on the back. "It's good to see you."

"Good to see you too."

"And this must be Naomi," says Deb as she releases me and grabs Nan.

"Call me Nan."

The long overdue reunion causes me to reminisce. Matt was always the smart one, the family prodigy, the firstborn child that always received so much attention, support and encouragement – things that were later rationed after the rest of us were born. Not that he didn't deserve it. He's so smart. He was valedictorian of his high school and graduated top in his class at Princeton University with a master's degree in Physics. He was a hard act to follow, and much of my adolescence was spent trying to be as good and as bright as my older brother. For me, an impossible task.

When Matt finally did fail at something – his first failure as best as I can tell, I celebrated. "He is human after all," I thought to myself, and I opened a bottle of expensive wine, and drank the whole thing by myself. At the time, it was a joyous day for me, because my own life was in ruin, and somehow, the thought of Matt being in pain eased my suffering.

After Matt graduated from Princeton, he landed a lucrative job with a defense department contractor, working on things so secretive that he was never able to discuss any details about his projects. For a while, he lived a lavish life, with money and prestige and cars and exotic vacations. But Matt was never happy in the corporate world and the corporate world became disenchanted with Matt. The politics and the competition and the stress and the pressures were difficult for Matt to handle and his work suffered. His performance reviews degraded over time, going from excellent to good to mediocre, and when Washington budget cuts began to affect his company, Matt was laid off. After nine years of service.

That's when I popped the cork.

The whole world now knew that Matt wasn't perfect, and I watched with sadistic joy as he went through several years of anguish, uncertainty and hopelessness – searching for a new direction in his life – just as I had. Matt worked various temporary jobs for some time, until a part time math and science teaching position at a small New Jersey college expanded into a full time job. The pay was not very good, but it was a stable, low pressure career. And since Deb was working, their combined income provided enough money to pay the bills and support a decent lifestyle. Matt has now been there for almost 20 years, and is head of the department.

The luggage finally arrives and we pick up Matt and Deb's suitcases and walk over to the car.

"Luke, you rented a minivan?" complains Deb.

"It's a long story. I'll tell you about it on the way."

We shove Matt and Deb's two suitcases in the back and pile the carry-ons on the floor between the two captain's chairs. Finally, we are ready, and we begin to leave the garage.

It is then that I hear it on the radio. My song. It is one of those songs you recognize as soon as the first chord is played. One of those songs you never grow tired of. One of those songs you know was written just for you.

A long, long time ago

It is one of those songs that is routinely butchered by semi-inebriated patrons in barroom karaoke competitions.

I can still remember

I immediately turn up the volume and wait for everyone's approving nod.

How that music used to make me smile

"God I hate that song," protests Deb from the back seat. "It just goes on and on and ooonnn – forever! It is soooo monotonous!"

I am insensitive to the Deb's objection. This is my song. This is my minivan, and I, the driver, am in control of the radio. So, I fumble with the buttons to add a little more bass and treble to the music.

The song is called "American Pie", but it is sometimes incorrectly referred to as "The Day the Music Died". It was written by Don McLean in 1971, and because McLean has never publicly explained the lyrics of the song, it has fostered decades of speculation about their meaning.

I can't remember if I cried
When I read about his widowed bride
But something touched me deep inside

The day the music died.

Almost everyone who has heard the song knows that "the day the music died" refers to the death of Buddy Holly on February 3, 1959. Holly, Ritchie Valens and The Big Bopper (JP Richardson) all died when the plane Buddy had hired to fly to a performance in Minnesota crashed.

But the rest of the song has been interpreted different ways by various people – including myself. My take on the lyrics is different than most. I'm not sure exactly why, but it is probably because many of my life experiences – or perhaps my life failures – seem to be reflected in the song.

Most people who have analyzed this song will tell you that it is primarily about the cultural changes that took place in the 1960's, but to me, there's a more personal meaning. To me, the song is a story about a young boy, growing up during a tumultuous time of change and redirection. A boy who has trouble understanding death. A boy who loses his faith in God.

It's the story of my life.

Bye, bye, Miss American Pie
Drove my Chevy to the levee
But the levee was dry

Some say the levee in the song was the name of a saloon. Others say the levee refers to a popular 1950's Chevrolet television commercial jingle – but my interpretation is different. My levee is a levee on the banks of a river, a river like the River Jordan – where people are baptized and saved from death – a place of repentance and salvation.

But the levee in this song is dry. No more water. No more baptizing. No more salvation. No more cheating death. "This'll be the day that I die."

There's a reason this song and its interpretation is so ingrained into my psyche. It is because I too, like Buddy's wife, had a tragic, unexpected death. I too, like Jack Flash in this song, lived a life in disarray, and I too, like the blue's singer, turned away. You see, I too experienced a day when my music died.

I was a lonely teenage broncin' buck
With a pink carnation and a pickup truck
But I knew I was out of luck
The day the music died

That one day changed my life forever. Why did it have to happen? I loved her so much. How could a loving God allow something like that to take place? How can you have faith in someone that lets you down like that?

That was the fateful day the music died for me. No more faith. No more hope. No more God.

Bye, bye, Miss American Pie

I am now singing out loud. "I do love this song", I manage to say between breaths. "Com'on, sing with me!"

Drove my Chevy to the levee
But the levee was dry

Nan joins in – half-heartedly. She's used to this behavior.

Them good ol' boys were drinkin' whiskey and rye
Singing "This'll be the day that I die
This'll be the day that I die"

About that time I realize I am in the "Return to Airport" lane instead of the "Exit Airport" lane. I guess I wasn't paying much attention to the road. I apply the brakes to the van but the breaks aren't very responsive and I am not able to change lanes in time. Too much weight I guess. Way too much luggage. But looping around Logan airport doesn't concern me too much – I'm singing.

Oh, and while the King was looking down
The Jester stole his thorny crown

It is widely accepted that the Jester in the song is Bob Dylan, and many think that the King in this verse is the "King of Rock and Roll", Elvis. However, my interpretation, again, is different. To me, the King is the "King of the Jews", Jesus. He's the rejected, defeated and crucified Christ, looking down from the cross. No longer able to carry the burden to lead his people or provide support. No longer able to help me in my greatest time of need. Not even able save the life of a small innocent child.

Bye, bye, Miss American Pie

About this time, I begin to hear a soft melodic baritone voice coming from the back seat. Matt too is singing along. How can he help it? The song is contagious. It draws you in.

Drove my Chevy to the levee
But the levee was dry

Matt's participation encourages Nan to sing more – and me to sing louder.

Them good ol' boys were drinkin' whiskey and rye

Deb just rolls her eyes. There is nothing she can do to stop this now.

Singing "This'll be the day that I die
This'll be the day that I die"

We are now circling past terminal E – again – and we get stuck in an onslaught of traffic. But it doesn't really matter, no one is paying attention.

'Cause the players tried to take the field
The marching band refused to yield

There has been a lot of controversy over who the "Players" are, but it's always been obvious to me. The Players are the players of the music, the proponents of faith and trust in God. They are the holdovers from the time when society and God worked together. When there was hope, and music, and dancing.

But the Players were not able to compete with the Sergeant Peppers Marching Band. Society had changed. God's influence is fading, and things will never be the same again. Bye, bye, Miss American Pie!

Bye, bye, Miss American Pie

By now, Matt, Nan and I are singing like a bunch of competing schoolgirls at a six grade coral concert, each trying to outdo the other.

Drove my Chevy to the levee
But the levee was dry

"Come'on Deb, join in!" I say to our still reluctant backseat passenger. "You know you want to. You can't sound any worse than we do."

"Good point," says Deb. "OK, OK. I give up. Here goes!"

Them good ol' boys were drinkin' whiskey and rye
Singing "This'll be the day that I die
This'll be the day that I die"

We finally clear the traffic and are headed out of the airport.

By now everyone is laughing, and everyone is singing. It is a cacophony of voices capable of scaring small children. It's great. Matt is slapping his hands on the armrests of his captain's chair to the beat of the music, and Nan is making a feeble attempt to add a little alto harmony. Deb is having trouble with the unfamiliar lyrics and begins making up her own words.

"Deb," I say, "It's the day the music 'died', not the day the music 'CRIED'."

"Well," responds Deb, "that falsetto of yours sure sounds like crying to me."

Bye, bye, Miss American Pie
Drove my Chevy to the levee
But the levee was dry

Matt reaches up and slaps me on the shoulder in a gesture of support for my singing. Our eyes connect for a moment as I look at him in the rear view mirror, and I realize that something is happening here that hasn't happened in a long time. My brother and I are beginning to enjoy being together – taking pleasure in each other's company. Something I think, both of us have desired for a while, but something that has been difficult for either of us to initiate.

It's funny how this song, a song that holds so much meaning for me, seems to be acting as a catalyst for bringing my brother and me closer together. I realize that no one else in the car really understands the meaning of the song's lyrics, and that no one knows how this song so aptly describes many experiences and tragedies in my own life – events that resulted in my ultimate loss of faith in God. But that's OK. What is more important now is that I sense that there may be some hope – a possibility that Matt and I may be able set things right – to mend a relationship that has been strained for so many years.

Them good ol' boys were drinkin' whiskey and rye
Singing "This'll be the day that I die
This'll be the day that I die"

The car screeches to a stop at a toll booth, but overshoots the toll attendant by about 3 feet. After backing up and paying the toll, I say, a bit embarrassed, "I think we may have to get these brakes checked. I don't think they are working very well." But no one is listening. It's the last verse. A verse that I know was written just for me.

And the three men I admire most
The Father, Son and the Holy Ghost
They caught the last train for the coast
The day the music died

"And now, the final chorus", I say, as we all join in, louder than ever.

Bye, bye, Miss American Pie
Drove my Chevy to the levee

But the levee was dry
Them good ol' boys were drinkin' whiskey and rye
Singing "This'll be the day that I ...

And at that moment, everything goes dark – and the music stops.

Everyone is startled for a moment, but then we all begin to laugh. We had all been so preoccupied with the song, that no one noticed that we were entering Sumner Tunnel – a narrow, dimly lit, carbon monoxide filled tube that connects Logan Airport to downtown Boston. The tunnel, which was constructed in 1934, travels beneath Boston Harbor, and since radio waves have difficulty penetrating water, mud and concrete, we had lost our signal. Nothing now but static on the radio.

"Pity," says Matt from the backseat, "looks like the music died – again."

"Maybe God is trying to tell us that he doesn't like our singing," jokes Deb.

But this is no time for joking, for at that moment fate intervenes once again in my life – the relentless mountain, conspiring against me.

Nan turns to face the rear of the van, to respond to Deb's comment, and doesn't see the impending danger. Maybe it is better that she doesn't know. There's nothing she can do – and nothing I can do. We are trapped in a sequence of events and circumstances that are beyond our control.

The traffic has stopped in front of us, and as I push on the brake peddle, to slow the van as it is gathering momentum traveling downhill toward the center of the tunnel, my foot hits the floor. No brakes.

"I knew there was something wrong with the brakes," I think to myself. "I should have stopped earlier."

I try to speak to alert everyone to our plight, but I can't. I am unable to utter a sound. Every muscle in my body has become paralyzed, and I am reduced into a helpless observer of the events about to unfold. I am almost catatonic, with both feet on the brake peddle, my arms stiff and my gaze straight ahead.

I don't know why I am unable to move. Maybe it's a hormonal or chemical imbalance in my body that overwhelms my nervous system in an emergency. Maybe it's simply my fear of death. I'm not sure. But it doesn't matter. I am frozen.

In front of us is a large tanker truck, sitting, stationary, in the middle of the lane ahead, waiting like a bumper block to stop a runaway train – and I realize that we are all about to die in a fiery crash.

The debilitating fear I am experiencing begins to affect my perceptions. Time seems to be moving slower and events are beginning to unfold is slow motion. The extra time allows my mind to wonder, and I start thinking about my daughter, and then about Nan. And I am overcome with sadness as I realize that I will never see Nan again. And I begin to face the harsh reality that our lives are over. Finished. And all the plans, hopes and dreams that we have for the future are gone – disintegrated in a few brief moments by a random twist of fate that has mercilessly intervened in our lives.

I wish, so badly, that I could believe that our souls, or our spirits, or our essence, or whatever ... would live on after our deaths. To believe that we continue some cognitive function after our bodies decay into dust. But that just seems too unrealistic. Not very logical at all. Just a false hope perpetuated by a world filled with religious zealots.

And now, my entire body is beginning to quiver as I confront the end of my existence. And as I sit there, helpless, watching as our van's momentum carries us closer and closer to the stopped truck, I begin to visualize our impending crash, as if it were already occurring. I see all of us dying an excruciating death as our van hits the 18 wheeler and the momentum carries our vehicle's chassis beneath the truck – as the trailer above slices our van in two. And then, I envision our decapitated bodies lying lifeless, in pools of blood on the fiery roadway.

Why me? Why is it that I was chosen to be put into this situation? A situation where I have no chance of success. No way out.

Where is God when I need Him the most? If he does even exist, which I doubt, He must be an empty, heartless soul. He has to know all the difficulties I have had in my life, and now He throws this at me. How detached. How unsympathetic. How cruel this God must be.

There is no hope. We are going downhill, picking up speed. We are goners.

Nan turns around, facing forward again, and sees the stopped truck. She screams, and instinctively braces for impact – and then helplessly, and hopelessly calls my name, "Luke."

But I can't look at her. I can't face the fact that I will be responsible for both our deaths, and I can't bear to say goodbye. So, I close my eyes, and wait.

But then, I hear a soft, calm voice coming from the back seat. It's Deb. She says, "It's OK Luke," with no panic or emotion in her tone. And very a-matter-of-factly she says, "Just use your emergency brake."

Such a simple solution. "Why hadn't I thought of that?" I think to myself. "But do I have enough time? And where is the brake? And how do I reach it, when my body is stiff and immovable?"

But the thought of losing Nan, of never seeing her again, causes me to fight against my paralysis. I force my hand off the steering wheel and to the right, and without looking, I locate a lever, in the center console between the seats ... and I pull ... and the van jerks to a stop – in plenty of time – a full car length, maybe two, behind the truck.

We sit there, silently for a moment. No one speaks.

Until finally, Matt pipes in, "Maybe we should sing a song," which causes us all to laugh.

The traffic has begun to move and the truck in front of us is now 30 or 40 feet ahead. The cars behind us are beginning to honk their horns, wondering why we are not moving. So, rather than holding up traffic, and waiting for assistance, and putting ourselves in another potentially dangerous situation, I begin driving forward, slowly, alternately releasing and pulling the emergency break to control our progress. Meanwhile, Nan finds the button for the emergency flashers, and turns them on.

We limp through the rest of the tunnel and exit into daylight. Because of construction there are concrete barriers on both sides of the roadway, and we must travel a while longer before we are able to exit, and pull over to the side of the road.

Then we all take a few deep breaths. We're safe.

Nan finds the rental car contract and calls the roadside assistance number on the back. The agency is very responsive and apologetic, and tells her they will have a tow truck bring us a replacement vehicle within an hour or an hour and a half. There are still no SUV's available, and now, no more mini-vans, so the company asks if they can send us a large 4x4 king cab pickup truck – with enough room for five passengers, and space for at least 20 suitcases in the back. Nan agrees without even asking for anyone else's opinion.

"Sure hope it doesn't rain," says Matt.

"Yeah. Me too," replies Nan.

The Coffee Shop

There is a small coffee shop a short distance from where we parked the van and we decide to wait there for our new truck to be delivered. Nan realizes that I am still shaken from the tunnel ordeal, and she takes my hand as we walk. She is so emotionally strong. So much stronger than me.

Matt and Deb seem to be fine, exhibiting no lasting effects from the brake failure episode. They are laughing as Matt tries to teach Deb some of the words to American Pie.

Seems like I am the only one who is still rattled by our brush with death.

Matt holds the coffee shop door open for everyone and I walk in behind Nan and Deb. Nan is the first to the counter, and orders some type of double something, extra swig of who knows what, sweet, latte, third world coffee-like stuff. And while she is deciding what size cup she wants, I turn to Deb and say, "Deb, I don't understand ..."

But before I am able to finish Deb interrupts, "You know, I don't understand either. This vocabulary is way above my head. Mam, I'll just have one of those brown icy things the lady over there in the corner has."

"That's not what I meant. I meant to say that I don't understand how ...", but at that moment the lady behind the counter gives me one of those are-you-ready-to-order-yet looks, and I say, "Large coffee," and then continue, "I don't understand how you ..."

"Bold or mild?"

"Bold! I don't understand ..."

"Room or no room?"

"No room!"

Matt then interjects, "Why don't you guys go get a seat and I'll pay for this."

"Thanks."

So, we wait a few minutes to get our drinks and Nan, Deb and I take a seat near the window. And then, I continue, hoping to complete my sentence before another interruption. "Deb, I don't understand how you were able to keep from panicking in the tunnel. When I heard your voice, when you told me to use the emergency brake, you were so calm. Weren't you afraid?"

"Oh," replies Deb nonchalantly, "I just didn't sense that we were in any real danger."

"So, you didn't see the truck?"

"No, I saw the truck. Even before Nan screamed."

"Then, you didn't realize that the brakes weren't working?"

"Oh no, I saw your face in the rearview mirror. I knew you were panicked, and it was pretty obvious the brakes had failed when you kept stomping your feet on the floorboard."

"I did?"

"Yeah, but I just had a feeling that we would all be OK."

About that time Matt sits down with his drink, and places four large cookies, each one different, in the center of the table. "Are you discussing your feelings again, Deb?" he says.

"Yeah, and I'm feeling like a cookie," says Deb as she lunges for the oatmeal-raisin.

"Wait a minute," I say, "That just doesn't make sense. There we were. No brakes. Gaining speed. About to plow into a truck – and you are 'feeling' that everything is going to be OK?"

"Yup."

"Doesn't make sense to me either," says Matt, "but I just accept the reality."

"What reality?" I ask.

"The reality that Deb's feelings are, more often than not – correct."

"But feelings are fleeting. You can't rely on feelings."

"I'm 'feeling' like grabbing that macadamia nut cookie, or maybe the chocolate chip," interjects Nan.

"No really. Don't you see what I am saying?"

"Yes", says Matt, "but you know, things are not always as they appear." You can't dismiss feelings as being non-factual or unfounded – just because you don't perceive or understand them."

"It still doesn't make sense."

"I didn't say it made sense," replies Matt as he reaches for a sugar cookie. "I just have learned to accept certain things that I don't completely understand – like feelings, and emotions. You see, it is very difficult to gather good scientific data on feelings because they are so hard to find."

"What do you mean?" I ask.

"Well, I have never been able to look at a 'feeling' under a microscope, and so, I don't really know what they are, and I don't know how they are generated, or how they are stored and remembered, or how they manifest themselves at different times, or how they are conveyed and shared with others.

"I don't even understand why I 'feel' like this cookie, which I have never had before, is going to taste so good," says Matt as he takes a huge bite. Then he pauses a moment, smacks his lips, smiles, and with his mouth still full, says, "I just accept the reality."

The Drive

It's not long before the rental car agency tow truck pulls up to our decommissioned van. They have already unloaded our new truck by the time we are able to walk back from the coffee shop. It's a big truck. Larger than we had imagined – a four wheel drive monster with oversized tires and jacked-up suspension that lifts the chassis a little over a foot off the ground.

The driver helps us transfer the luggage to the back of the truck and gives me the keys. I look at them for a moment, and then I look over to Matt. “You wanna drive?”

“Sure,” says Matt, “I've always wanted to drive one of these things.”

I give Deb a boost into the front seat, and the rest of us climb in. And we are off, once again.

We merge onto interstate 93 and head north across the Bunker Hill Bridge and past Paul Revere Park. The hum of the large truck tires on the highway is almost hypnotizing, and I am just about to fall asleep when Deb turns around and says, “Luke, tell me about your problem.”

I look up, puzzled for a moment, not fully understanding the question.

“I've got a list of problems, if you'd like to look at it,” pipes in Nan.

“No, I think you know what I mean Luke. In the tunnel. You were so overwhelmed with fear. Why do you think you reacted the way you did?”

I was hoping no one would bring up the subject, and just forget about what happened in the tunnel. Leave it in the past. But somehow I knew, in the back of my mind, that Deb wasn't going to let the memory of today's events fade away, and I could tell from her question that she thought that I needed help. And I knew Deb's training was going to kick in, and she was going to pursue this discussion – a discussion I desperately want to avoid.

Deb works as a counselor for a group of psychologists in southern New Jersey, and knows how to identify emotional problems and disorders. And she doesn't shy away from confronting them head on. It's what she does. Every day.

So, I know what I have to do. Somehow, I need to deflect Deb's question – make it appear that I responded normally in the tunnel – just as anyone else would have. I need to answer in a manner makes it appear that Deb Is being over reactive and that her question is inconsequential. And then, I need to change the subject.

I give it my best try. "You know, I don't feel that my reaction to that particular situation was that abnormal. Most people panic to some degree in emergency situations. I honestly thought we were going to be seriously injured in that tunnel, and I think my response was pretty normal – the same type of response many people would have exhibited under the same circumstances. Matt, I think we may be losing our radio station signal. Do you think you can try to tune in another?"

Unfortunately, Deb wasn't deterred. "Luke, I think you know that there's more to it than that. There's something deeper."

"How do you know that?"

"I just sense it."

"Your feelings again?"

"Yeah."

Then, Nan turns toward me, and very softly and compassionately says, "You know, it may help you to talk about it."

I pause for a moment. It's not easy for me to talk about myself. Especially about my faults and shortcomings

– and about my feelings. It’s much easier to keep everything inside – out of public view. It's safer that way.

I learned this defensive behavior while growing up. My Dad had a very strong and domineering personality. It was best never to cross him. He was a “my-way-or-the-highway” type of guy, and anytime I expressed an opinion, or an emotion, or an idea that was different from his, I was attacked. Not physically. He never hurt any of his kids, but he had a deep, forceful, resounding voice, that could scare the living daylights out of an adolescent child, and whenever anyone in the family objected to Dad’s standards, or views or opinions, he would set us straight.

And so, over time, I learned to keep my feelings and my thoughts to myself. I discovered that the best way to avoid confrontation with my Dad was to avoid any interaction. That worked pretty well, but as I grew older, the behavior that saved me from my father's wrath was so ingrained into my personality – that it became a defense mechanism for dealing with other people as well. And I began building that island around myself – the island Nan said I needed to abandon. The island she has been trying to build a bridge to – for almost 7 years.

“Luke, Luke, anybody home?”

“Sorry Deb,” I say as my focus returns. “I just needed a minute to compose my thoughts. It's not easy for me to talk about.”

“I understand. Take your time.”

“Well, I guess it all began back when I got cancer. My fear of dying, that is. In one day, I go from being a strong, healthy, self-sufficient, invincible, middle-aged man – to a guy who is one-slip-on-a-banana-peel away from the graveyard. And that was hard for me. Still is hard.”

'You know,” says Matt, “We're all going to die.”

“Yeah, but you don't think about it, or expect it to happen until you are old – until you are wasting away, listless, in a nursing home bed, with needles and tubes and drugs and enough dementia to numb the reality of the situation. It's much easier to accept death when dying becomes a better alternative to living. It's much harder to accept it when you feel like your best years are ahead of you."

"How's your cancer doing?" asks Deb.

"Oh, pretty good. It's been in remission now for 7 years. They say I am over the hump. Apparently, if you get past 5 years, your chances are pretty good. I still have regular scans – to make sure there are no new growths, but so far everything is good."

"He just had a scan before we left," injects Nan. "We should get the results sometime this week."

"How old were your parents when they died?" asks Deb.

"Oh, I don't really remember. Somewhere in their early 50's."

"It was a car accident?"

"Yeah, it was raining pretty hard, and Dad lost control of the car and hit a tree."

"Do you think that your parents' deaths are somehow related to your fear of dying, and to your paralysis in the car?"

Now I am becoming uncomfortable. That is a tough question for me to answer. I know my parents' deaths have left an emotional void in my life. I had never been able to reconcile with my father. We never worked out our differences – never had that moment when we hugged and said that we were sorry – and we were never able to say, "I love you" to each other.

But I am not about to bear my soul in a 4 by 4 pickup truck on Interstate 93, and I am not about to diverge any suppressed feelings from the past. So, I answer Deb's question just like I had answered others many times before.

"No, not really. At the time, my parents' deaths weren't that traumatic for me. I had so many other problems at the time – losing my job – my divorce. The accident was a very sad and tragic, but I'm not sure I ever cried. I was never really very close to my parents – especially my father, and somehow I was able to accept their deaths. I'm not sure why. Maybe it had something to do with various childhood scars that have not fully healed – but I think it was probably just because I was so focused on my own troubles."

"I understand."

"Yeah."

"I know it is probably hard for you to talk about Luke, but what about your daughter?"

That's enough. I am finished talking. I was already out of my comfort zone and now, the conversation is headed somewhere I do not want to go. So, after a short pause I say, softly but deliberately, "I can't talk about that right now."

"That's fine. I understand," responds Deb unexpectedly. "Just let me know if you ever feel like discussing it." And then she abruptly drops the subject. "Matt, you find that radio station yet?"

It's not long before we cross the Massachusetts – New Hampshire state line, and we make an obligatory stop at the New Hampshire State Liquor Store just off Interstate 93. We enter the store and all head in different directions. Matt makes a beeline for the liquor section while Deb homes in on the mixers. Nan and I head for the wine.

We all meet at the checkout counter a few minutes later with two shopping carts crammed full with an eclectic mix of international booze. Matt has assembled a formidable collection of spirits that would make a top shelf bartender proud. He has vodka from Russia, scotch from

Scotland, gin from London, tequila from Mexico, rum from Jamaica and bourbon from Kentucky – and Deb has piled on an assorted collection of everything you could ever dream of mixing with anything 80 proof.

Meanwhile, Nan and I have embarked on a worldwide wine tasting tour, selecting a bottle of red and a bottle of white wine from France, and from Italy, and from Spain, and from Argentina, and from Chile ... and from Australia – plus another nine or ten bottles of some of our favorite California vineyards.

"Who's gonna drink all this stuff?" asks Deb.

"I don't know," says Nan, "but I'm glad we have a big truck."

The checkout guy boxes all the bottles into 6 empty wine and liquor cartons and helps us hoist everything into the back of our truck. Then we jump back into the cab, merge back onto the interstate and continue our journey.

Matt finally finds a new radio station and turns up the volume. They are playing an old song by a group called Steam. It's a song they often play at sporting events, called, "Na Na Hey Hey Kiss Him Goodbye." Everyone joins in.

Na na na na
Na na na na
Hey hey hey, Goodbye.

I turn to Nan and begin singing the first verse to her.

He'll never love you
The way that I love you

Nan smiles at me briefly, but then turns away and peers out the side window. It's a reaction I have grown accustom to lately. It seems that whenever I begin to express my affection toward Nan, she becomes distracted and aloof. It's as if she is not interested in anything I am saying. And this has been happening with greater frequency lately, and I have become confused – not knowing how to translate her body language into a dialect that I can understand.

I reach over and gently put my hand on Nan's shoulder, and she begins to turn toward me, but halfway around she stops – and screams – "Aaaaaah!"

"Nan. What's wrong?"

"There's a bee!"

I look over and there's a small yellow jacket buzzing against Nan's window. The poor frightened fellow must have inadvertently flown into the truck when we stopped at the liquor store.

Bees typically don't scare me, but Nan is deathly afraid of all insects – especially the ones that sting, and I know that

I have to act quickly to avert a bad situation. So, I grab a newspaper from one of the carry-ons on the floor, roll it up, and I say, right in time with the music, *"Na na na na, Na na na na, Hey hey hey, Goodbye"*. Then I take a powerful swat at the pest. Unfortunately, the top of the paper hits the handle on top of the door and diverts my swing to the right, and I miss the bee completely. The attempted assassination apparently perturbs the little fellow and he begins flying wildly around the back seat. I try to swat the bug in mid-air but after three or four failed attempts I accidentally elbow Nan in the head.

"You OK?" I ask. "I'm so sorry."

"I'm fine. Just get that sucker!"

While I am attending to Nan, the bee flies to the front of the truck in an attempt to involve the whole family in this life or death battle.

Deb immediately joins the confrontation by swinging her purse back and forth through the air, and Matt grabs the newspaper from me and attempts to squash the bee against the windshield. In the meantime, the truck is swerving back and forth on the interstate as Matt is trying to multitask as both driver and exterminator.

By now the bee is beginning to wear down and lands on the dashboard, apparently gasping for breath. "We got you now," says Matt. *"Hey hey hey, Goodbye."* But as Matt attempts a final, fatal blow with the newspaper, Deb, at just about the same time, swings her purse and hits Matt's arm, which causes the truck to veer to the right and throw everyone to the left. Matt pulls the truck back into our lane but the diversion gives the bee a reprieve. And he takes off, headed once again for the back seat, straight toward my head.

I duck, and the bee flies by and lands on my window. And I cautiously reach over and push the power window button. The window opens, and the bee flies away.

So kiss him
(I wanna see you kiss him)
Go on and kiss him
Goodbye

"That was exciting," says Deb.

"I think the excitement may just be starting," responds Matt. "Look behind us."

We all turn around and see flashing blue lights. Not a good thing.

"Looks like the New Hampshire State Highway Patrol," says Nan. "How fast were you going Matt?"

"I'm not sure. I was paying more attention to the bee than the road. Anyone know what the speed limit is here?"

No one answers as we slow down and pull over to the side of the road. Matt pulls out his license and Nan hands him the rental car contract. The state trooper pulls up behind us, but stays in his car and talks on his radio for an intimidating three minutes before he slowly walks up to Matt. Matt rolls down his window and hands the officer his license and the contract without saying anything.

The trooper is the first to talk. "You seem to have been weaving all over the road."

"Yes sir," responds Matt. "We were trying to kill this bee in the truck."

"I see. Where is this bee?"

"He flew away."

"I see. You know," he says while peering into the back of our truck, "there appears to be a lot of alcohol in the back here."

"Yeah ... ah ... we're going to the lake region for a week, and we just picked up a few things at the state store."

"I see. You're going to drink all of this in only one week?"

"Yeah ... ah ... we might have gotten a little carried away."

“I see,” says the trooper as he looks over the rental car contract. “You know, this contract is for a minivan. Not a pickup truck.”

It wasn’t until that moment that we realize that we did not receive a new rental agreement from the car company when we swapped out vehicles. Matt tries his best to explain. “Well, our minivan broke down in the tunnel, and this is a replacement.”

“I see.”

“We can probably call the rental car company for verification.”

“Do you know how fast you were going?”

By now, it was obvious to everyone in the truck that this conversation is not going well. Matt has done a valiant job so far answering some very difficult questions, but I have always had trouble responding to the “Do you know how fast you were going?” question. What is the best way to answer? Should you tell the truth and admit you were speeding, or should you lie and say that you were going the speed limit, or should you say you don’t know and admit that you were not paying attention to your driving. I am curious to see which one of these options my smart brother is going to choose. But what he says surprises me. I think it surprises the trooper as well.

“Well,” says Matt, “... you know ... it really depends upon your point of reference.”

“Your point of reference?”

“Yeah, you see, to the bee that was flying around inside our cab, we weren’t moving at all. From his point of reference, he was attacking a group of people in a car that was perfectly still. The only motion he was aware of was the Deb's swinging purse.

“But, from the point of reference of someone with a radar gun sitting on the side of this highway, we were traveling, say, 55 miles an hour.”

The "55 miles an hour" statement was a bit of a stretch, we were obviously going a lot faster than that, but Matt doesn't miss a beat. He continues.

"Now, an observer floating above the earth would have seen our car moving some combination of our car speed, plus the speed of the earth's rotation. That would result in a speed probably somewhere around 800 miles per hour. You see it all depends upon your point of reference – where you are viewing the events from."

Matt was on a roll. This was a teaching moment for him. He was asked a question, and he was going to answer it just like he would answer any student in his college classes – with a sound, accurate, and complete scientific explanation. He continues.

"An observer on the sun would see our car traveling some combination of our car speed, the rotational speed of the earth and the 67,000 miles per hour the earth is traveling around the sun. And ... an observer in the center of the Milky Way would see our car traveling an additional 500,000 miles an hour as the sun circles the galaxy. So, you see, speed is almost always dependent on your point of reference."

"I see," replies the officer as he tilts his head slightly to the side and presses his lips together. He then asks Matt to step out of the truck – so that he can conduct a sobriety test. Matt complies and walks a straight line with no trouble. The trooper then tells Matt to wait in the truck while he verifies our replacement rental car story. After a few minutes he returns and says, "OK. You can go now. Drive safely. And please pay attention to the road."

"Thank you," says Matt, obviously relieved that we were not going to get a ticket. And that should have been the end of the conversation. But Matt wasn't quite finished. He adds, "Oh, just one more thing."

"Yes."

"You know, there is one exception. There's something really odd about light. You see, light travels at the same speed regardless of your point of reference. Even if you are moving incredibly fast, and traveling toward light, or moving equally fast traveling away from light. When you measure the speed of light, it will always be the same. It doesn't matter whether you are running toward it or away from it. It's always the same speed."

"I see. Have a good day."

We merge back onto interstate 93 and after we are a safe distance from the patrol car, Matt says, “I can't believe he didn't give us a ticket.”

“I think it was because you confused him,” I responded.

“I'm sure you did. You confused me,” added Nan. “You know, I followed the part about the rotation of the earth, and about circling the sun, and speeding around the milky way ... but I'm not sure I understood that light thing.”

“Well, you are not alone. Most people have trouble with that concept, and almost everyone I talk to has trouble understanding the implications.”

“What implications?” asks Nan.

“Well, do you have a few minutes?”

“Sure. We're not doing anything else.”

“I think I may take a nap,” says Deb. “I've heard all this before. Matt talks about this stuff all the time – even at cocktail parties and dinner engagements. I think it's the main reason we don't have many friends.”

“I don't believe that,” I pipe in. “I think this is interesting.”

“That's because you're related.”

“It's OK Matt. Go ahead.”

“Well,” begins Matt, “Luke, do you remember that bee that was flying straight toward your head?”

“Oh yeah.”

“So, relative to you, how fast do you think he was flying?”

“Oh, maybe five miles an hour.”

“OK, now, you saw the bee coming toward your head. How were you able to see it? What was happening?”

“I'm not sure I understand the question.”

“Well, let me explain, the reason you were able to see the bee flying toward you – was because light was reflecting off the bee's head, and traveling through the air to the retinas in the back of your eyes."

"That's pretty obvious," says Nan.

"Yes, but now here's the critical question. How fast was that light, the light reflecting off the bee's head, traveling toward Luke's eyes? We know the bee is flying about 5 miles an hour toward Luke and we know light travels at the speed of light."

"OK," replies Nan, "You would think the light would be moving at the speed of light plus 5 miles an hour. But I guess that's wrong. Hugh?"

"Yes, that's right. It's wrong. Light always travels at the speed of light. Its speed is independent of any other motion. It doesn't matter if you are chasing after a beam of light or running away from one. Regardless of any motion, light's speed is constant."

"But why does that even matter? The bee is flying pretty slow, and light travels at a million miles an hour."

"Actually, it's about 670 million miles an hour. But you have a good point. In this case, where the bee is moving relatively slow, it doesn't matter. And it doesn't matter for many of the other experiences we typically have during the day. But you see, things are not always as they appear. Our experiences are limited by our senses which only allow us to see a very small portion of the world we live in. And when we probe beyond our senses, we find a much stranger world. A world you would not recognize.

"For example, imagine our bee friend jumps on a rocket ship traveling three quarters the speed of light towards Luke, and Luke hitches a ride on another rocket traveling three quarters the speed of light towards the bee. Now, how fast does Luke see that bee coming toward his head now?"

"I know this is probably wrong," says Nan, "but you would think that Luke would see the bee coming toward him at the bee's speed, plus Luke's speed. Three quarters the speed of light plus three quarters the speed of light. One and a half times the speed of light."

"You would think that, but that's not the case. The light reflecting off the bee's rocket ship and reaching Luke's eyes can only travel at the speed of light. No faster. Therefore, it's impossible for Luke to see anything coming toward him traveling faster than the speed of light. It just can't happen. Light always travels at light speed. And, because of that, there is no way for Luke and the bee to be able to approach each other any faster than the speed of light – no matter how fast they are traveling toward each other"

Now I am confused. "That just doesn't make sense," I say. "Don't the people in Star Trek travel faster than the speed of light? You know warp 3, warp 4."

"That's why they call it science fiction. It's fiction. It can't happen. Nothing can go faster than light."

"But if both the bee and I are traveling toward each other at three quarters the speed of light, and we can't close in on each other any faster than the speed of light ... then something else has to give. Something has to compensate for the discrepancy between the speedometers on our rockets and the speed we are approaching each other."

"Good point," exclaims Matt. That's the same conclusion Albert came to."

"Who's Albert?"

"You know, Albert Einstein."

"Sorry, I didn't know you two were on a first name basis. So, what did Albert say? What gives?"

"Well, Albert realized that when you are traveling at very fast speeds relative to others watching you, that something strange happens. Your watch starts to malfunction. It ticks slower than normal, like the battery is wearing out. And it's not just your watch, you move slower, and you breathe slower, and your heart beats slower. You see, the faster you travel – the slower time passes. That's what gives. Time."

"But that just doesn't make sense," objects Nan. "Time is just time. It's not a something. It's just an interval between two different events. It's just how long it takes. It's not something you can stretch out. It's not an object. It's just a measurement. You know?"

"Yeah, I know that that is what lots of people think."

"But probably not Albert? Huh?"

"No. Albert said that time is not a constant. Time gets longer and time gets shorter, and two individuals in relative motion will measure time differently, and they will disagree about the time events occur."

"And that's all because nothing can go faster than the speed of light?"

"Yup."

"So, why should we believe Einstein? Maybe he just made a miscalculation. Even geniuses make mistakes. Don't they?"

"Well, you know, when Einstein first proposed his theory of special relativity in 1906, there were more skeptics than believers."

"And now?"

"Now, no one in the scientific community doubts Einstein's concept of time. Whenever someone is in motion, relative to you, their clocks tick at a different speed than yours. It's been proven over and over. We don't notice it because in most cases the differences are too small to measure, but when you increase your speed the effect becomes dramatic. Satellites circling the earth have to have their onboard clocks adjusted because they run at different speeds than clocks on the ground."

"Really?"

"Yeah, although the difference between the two clocks is still very small – imperceptible without very accurate instruments. But the time differences become dramatic if you are able to accelerate to extremely fast speeds. For instance, let's say a 30 year old astronaut leaves behind his 10 year old son on earth and travels on a spaceship to a distant star at 95% the speed of light. When he returns to earth 30 years later, he will have only aged 10 years."

"So the astronaut and his son would both be the same age? 40?"

"Yup."

"That's hard to believe."

"I know. It goes against your gut. Against your experience. But you see, things are not always as they appear. The reality we experience day to day is not the reality of creation. The time that we measure here on earth is not the time of the universe."

I pause for a moment, not quite understanding exactly what Matt is saying. How could time possibly be affected by speed? And why does light have some arbitrary speed limit? I'm a logical guy that lives is a world that can be logically explained, and this new concept of time just doesn't fit into my concept of reality. If time is measured differently for two different people, then everything they see and observe will be out of kilter. You can't have one single event happening at two different times. It just can't work.

"I still don't understand how time can be shorter for one person than for another," says Nan.

"I know it's a difficult concept. But you have to keep going back to the universally accepted principle that light always travels at the speed of light. And because of that, time has to slow significantly when two objects travel at high speeds relative to each other."

"I still don't get it."

"Well, just give it some time to sink in, and after a while it begins to make sense."

"OK."

"You see Nan, you were wrong earlier when you said that time is not something. Time really is something – just like space is something. And time stretches and shrinks – just like space stretches and shrinks. Albert took both the concept of space and the concept of time and put them together, because they are so intimately linked with each other, and he showed how time and space can expand and contract together. He called it, obviously, space-time."

"Ah ... now I'm really confused. Space stretches and shrinks?"

"Sorry, too much information. Class is over for today. We can talk about this more later – if you are interested."

"OK," says Nan. "You interested Luke?"

"Yeah," I say half-heartedly – not being quite sure that I want to endure a whole week of lecturing from my know-it-all brother.

"What about you Deb? You interested? ... Deb? ... Deb? ... Wake up Deb!"

We exit Interstate 93, drive through the quaint little New England town of Ashland, and follow Daniel Webster Highway to Holderness, a small, charming, and picturesque municipality that sits on the southern shore of Squam Lake. Since all we had had for lunch was a cookie at the coffee shop, we stop for a few minutes at the Squam Lakeside Café for a mid-afternoon snack, where we split two lobster rolls. So good. The café is down the street from the Holderness Fire Department, and the standing joke is that if your house catches on fire, you will probably have to wait for the firefighters to finish their lobster rolls before they respond to your emergency – and pray that they don't go back for seconds – or decide order some homemade ice cream for dessert.

After about a half hour, we continue our journey and turn left onto Squam Lake Road which circles around the lake. We then drive five or six miles and turn right onto the dirt and gravel road leading into Deephaven Camp.

"Hasn't changed much. Has it?" I ask Matt.

"Nope. Looks exactly the same to me."

We pass the tennis courts, and the ball field, and the dining hall, and pull over to the side of the road and park the truck 30 or 40 feet from our cabin – as close as we can get. There's another car parked next to us with Massachusetts plates.

"Looks like Mark, John and Mary may have beat us here," says Deb.

Mark, John and Mary are my other three siblings. Mark flew in from Dallas, John from San Francisco and Mary from Atlanta. They met in Boston, as we did, and drove up together.

"Guess they didn't have any car trouble," says Matt.

Everyone jumps out of the truck, but I wait for a moment – not sure if I want to get out. It's been so long since this family has been together. Five kids. All so different. Nothing in common. Each gone his separate way. Maybe getting back together is not such a good idea. Who wants to dredge up the conflicts of the past – start that all over again? We have all grown older, and hopefully more mature, but we have also settled into lifestyles that are completely different from one another. Family and heritage can be a strong bonding force, but divergent vocations, interests and attitudes can overpower ancestry. It's like trying to hold together the positive ends of two bar magnets. You can do it for a while, but as soon as you let go, as soon as you stop forcing them together – they repel. They will never stay together, and there is nothing that the magnets can do about it. They separate because of what they are and how they were made. No coaxing, or good intentions or therapy is going to be able to hold them together. Nature has dictated the outcome.

I fear that something similar is going to happen here.

Deb is the first to begin walking down the narrow, rocky path to the cabin. Matt and Nan are close behind, and I, lagging behind a bit, bring up the rear. But before we get too far, the screen door on the front porch swings open and Mark, John and Mary come running out to greet us. Deb has met everyone before, the last time being at our parent's funeral, and she performs her standard bear hug routine with a few relatively loud woo-eee's. Her exuberance is contagious, and the first act of this reunion goes flawlessly – with hugs, and laughter, and pats on the back, and to my surprise, a sincere and joyful feeling of being together again. It's as if we are all great friends, who actually missed each other during the years we have been separated. It's a good feeling, and I sense that this may be a new beginning for our family. Maybe there is hope. Maybe, for once, harmony will prevail. I guess I should not have been so worried.

But then again, it may be a long week.

After a few minutes of "How ya doing?" and "Ya look great for your age!" and "How was your trip?", the girls, Nan, Deb and Mary decide to walk down to the dock to view the lake up close, and the four male siblings, Matt, Mark, John and I, start unloading the luggage.

"Where did all this come from?" asks John as he peers into the back of the truck.

"I'll explain later."

"OK. I'll set up the bar," says Mark. "You guys can handle the suitcases."

"OK."

We transfer all the bags and boxes from the truck bed onto the grass and begin carrying everything one by one to the cabin. When I open the screen door with Nan's first suitcase, and walk past the sink and the icebox and on into the living room – I feel as if I am traveling back in time. So many memories – the squeaking sound of the spring pulling the screen door closed behind me – the slightly musty smell of the 100 year old floorboards and the cane back rockers that have survived countless harsh New England winters – the cool damp breeze from the lake coming through the open windows. There is a familiarity and a certain tranquility that reminds me of times past. Of family times together in a quaint cabin on a serene lake – where family squabbles and personality conflicts seem to be trumped by the beauty, majesty and power of a very unique and special place – a place like nowhere else in the world.

The cabin is just as I remember it. The old brick fireplace in the living room, the huge screened in porch overlooking the lake, and the windows with sash cords that open completely down into the wall. There are two single bedrooms downstairs and two doubles and a single upstairs – all with only twin beds, making it very difficult, though not impossible, for two people to sleep together. There are only two showers, and two toilets, all in separate 4 by 4 square foot rooms. Not a luxury hotel by any means. But luxury hotels don't feel like home. This does.

The staircase leading upstairs is so narrow that it is almost impossible to carry a suitcase on your right or left side, so I lift Nan's 65 pound monster in front of me and begin ascending the small creaking steps. About half way up I become a bit dizzy and have to sit for a minute.

"I would help you," says Mark from below, "but I have to get this bar set up."

"That's OK, I've got it."

I then turn around on the stairs, in front of the suitcase, and begin pulling it up one step at a time, briefly sitting on each step as I struggle to the top.

Then I repeat the process for the second suitcase, after which I collapse in the rocking chair in front of the fireplace to rest.

"Just relax for a few minutes", says Matt. "John and I will get the rest."

"Thanks."

"Wanna drink?" asks Mark.

"Just some water right now."

About that time Mary and Nan walk in the back door. They seem to be having an argument about something, but I am not able to hear the conversation.

Mark hands me a drink in a red plastic cup. "Here's your water," he says. "Oh, by the way, I added a splash of scotch to it."

"Good, I think I may need it. Is there a problem Nan?"

"No!" she says, obviously annoyed.

"Mary, what's going on?" I ask.

"Oh, nothing really. I was just ah ... wondering, if it was really ah ... "

"She thinks that it's not appropriate to have a secular wedding in a church!" interrupts Nan.

"It's not a church!" I say trying to hold back my anger. "It's an outdoor chapel!"

"It's still the house of the Lord!" shouts Mary.

I take a deep breath, trying to control my temper, but I know my blood pressure is rising because I can feel a pulse in my neck. I was afraid that Mary was going to react this way to the wedding, given her fundamentalist Christian values, but I had hoped that she would keep her opinions to herself. I guess I was wrong.

"Here Luke," says Mark as he hands me another red plastic cup, "I made you another drink. Less water this time."

"Thanks. I may need another in a few minutes."

"No problem."

No one says anything else for a moment, apparently wondering if this argument, just like so many arguments in the past, is going to escalate into a full blown family feud, and we all begin nervously looking back and forth around the room at each other.

"Well," says John, finally breaking the silence, "I know it is going to be a bit of an unusual wedding, but I think we can all agree that marriage is a wonderful thing, and that this week should be a joyous family time."

We all know, of course, that John is right, but he may be understating the actual problem. This is going to be more than just a 'bit of an unusual wedding'. This wedding has morphed from the union of two young people in love, into a worldwide cultural conflict. A Middle East crisis. That was the main reason we decided to have the celebration in a remote part of New Hampshire at a small outdoor chapel, on an island in the middle of a lake – far enough away that it would be easy for anyone to decline an obligatory invitation.

We all had hoped that Nan's daughter Ruth and her betrothed would have eloped – spent a weekend in Vegas and consummated the deal in private. But they wanted a public wedding, and after months of haggling and arguing, and making and canceling plans – this was the compromise.

I, personally, do not have any problems with the wedding. Religious and cultural boundaries have never been very important for me. After all, I married a Jewish woman, and although Nan's family didn't approve, time has ameliorated many of the tensions and hurt feelings.

But this is different. Ruth is marrying a Muslim.

And that announcement did not sit well with any of Nan's immediate relatives.

Nan was raised in a very traditional orthodox Jewish family, and although she is not as conservative or observant as her parents, her families' approval is still very important to her. That was not always the case. Nan rebelled against her parents' strict beliefs in her late teens, and there where many years where she had little contact with any of her immediate relatives. It was only after she married and had a baby that she began seeing her parents regularly. But that may not have been a good thing, because the tension caused by family conflicts was one of the contributing factors leading to Nan's divorce.

And so, from the time Ruth, Nan's daughter, was about 2 years old, Nan raised her as a single parent, and although she intended to bring up Ruth with traditional Jewish beliefs, as Nan's parents had insisted, the demands of a single mom with a full time job didn't leave much time for any formal religious training.

Ruth always knew she was Jewish, but she was really only Jewish because of heritage and culture, not because of her faith. And therefore, it was easy for her to live a secular lifestyle and to nurture relationships with non-Jewish friends – and to fall in love with Abul, who she met at the University of Chicago.

Not long after Ruth left for college, Nan's parents confronted Nan and accused her of being a failure as a parent. That hurt her very much, and she countered their accusation by telling them that their parenting skills were far worse than hers. The conversation then deteriorated into a tirade of slurs and barbs – fueled by almost 40 years of pent up feelings. The argument ended, finally, with a broken vase and a slammed door.

It wasn't long after that – that Nan and I were married, and I have always wondered if part of the reason that Nan accepted my proposal, was to retaliate against her parents. I may never know. But I do know that our marriage served as an example to Ruth, and gave her the courage to accept Abul's proposal.

Funny how things work out. How every decision we make and everything we say affects others in ways we don't anticipate.

"Sorry," says Mary, "I shouldn't have said anything. I didn't mean any harm."

"It's OK," replies Nan.

"I guess I need to be a little more open minded."

"Where's Deb?" asks Matt, changing the subject.

"I don't know. We left her on the dock a few minutes ago," says Nan.

We all walk out onto the back porch overlooking the lake and our dock. John is the first to see Deb. She is standing on the steps outside the back screen door, obviously in distress. Her breathing is fast and shallow and she is holding onto the railing with both hands, apparently having difficulty standing up.

"Deb. What's wrong? Are you OK?" asks John.

Deb doesn't respond, and Matt rushes out the door and helps Deb to sit down on the steps. "Just take deep breaths," he says. "You'll be alright."

After a few minutes Deb appears to be a little better. She is breathing normally and the color in her face has improved. "Why don't we go inside and you can lay down," I say.

"No, no, I can't go inside. I can't," says Deb as her breathing again becomes erratic and labored.

"Why not?"

And then something totally unexpected happens. Deb turns her head and looks directly into my eyes and says in a shaky but deliberate voice, "Because we're not supposed to be in there."

"What?" "What do you mean? Why aren't we supposed to be in there?"

"I don't know," she responds timidly, after a brief pause.

"What do you mean you don't know?"

And again, in that same haunting voice, she says, "We're not supposed to be in there."

For a moment, I am speechless. I had known for a long time that Deb had certain personality "irregularities", as Matt once explained to me. Many years ago she had a period of deep depression and spent a number of months in counseling. Matt said at the time that the therapy had been successful and that the depression was cured, but a few years later Deb had to return to counseling for some reason which was left unexplained. Most of the time Deb seems to be a perfectly normal, well balanced, fun-to-be-around lady, but then something like this happens and you begin to wonder if her head is screwed on properly.

"Matt, do you have any idea what she is she talking about?" I ask.

"I'm not sure," responds Matt. "But I'm not going back in the cabin."

"Why not?"

"Didn't you hear? We're not supposed to be in there."

I am now screaming, "WHY aren't we supposed to be in there?"

"I'm not sure," repeats Matt. "But I'm not going back in the cabin."

"Matt, I don't understand. Why are you afraid?"

"Because we're not supposed to ..."

"This is ridiculous. Matt, you're a physicist for God's sake. Give me a sane, rational, scientific reason why we can't go back inside."

"I can't. I wish I could, but I can't. All I know is that when Deb senses danger, there is a high probability that there is a problem. And I have learned over time, to trust her instincts."

"That's just stupid!"

"Just because you don't understand something doesn't mean that it's stupid. It might just mean that you don't have the knowledge or the skill to interpret it."

I turn around and look at everyone else, expecting words of support from my other siblings, but everyone is quiet. Dazed I guess – not sure what to think. John seems to be a little fearful, Mark is chugging down the rest of his scotch and Nan just shrugs her shoulders when I look at her. No one says anything for a moment, until I feel compelled to break the awkward silence. "Come on guys. Let's be realistic. This is our cabin. We've always stayed here ... and we've never had a problem before."

"Maybe something is different or changed or something," says Nan.

"I can't believe this. Am I the only rational person here?"

No one answers my question, and we just stare at each other for a moment, until we hear a rustling in the bushes behind us, which startles everyone, and almost causes Nan to fall off the porch steps.

"Excuse me," says a portly gentleman who walks out of the woods and up to our porch steps. And as Nan regains her balance, he continues, "I apologize for the interruption. My name is Stephen – from the cabin next door. And I overheard you talking, and ... ah ... you know ... we wouldn't mind trading cabins with you. You see, you have two showers in this cabin, and we only have one shower and this old bathtub with legs on it. And that's a problem for us because we have five teenagers staying with us who spend an inordinate amount of time in the bathroom. And we could really use an extra shower."

"You see?" I reply to everyone. "This guy isn't afraid of staying here. Why should we be?"

"Maybe Deb feels something – something we can't perceive. Maybe there's a good reason she's afraid," says John.

"That makes no sense at all. Every time you cross the street you can perceive that there's danger. Every time you get into a car. Every time you ride a roller coaster. There's always danger, but you can't live your life like that. Otherwise, you would never get out of bed in the morning."

"Yeah, but I think that this is different."

"Oh, come on. There is nothing wrong with this cabin. There are no spooks in there. Let's just go inside."

"Deb," says Matt, "Are you sure about this?"

"I can't stay in that house," she replies.

"I'm not sure I want to go back in there myself," says John.

"But this is our cabin," I plead. "Staying next door just wouldn't be the same."

"Oh I don't know," says Mark. "It's the same view, same type of cabin, same wide screen TV."

"There is no television!"

"Precisely!"

"What do you think Nan?" asks Mark.

"Well, she responds, I know this cabin has sentimental value for you guys, but you know, Mary, Deb and I have never been here before, so I don't think it really matters to us."

"No, it doesn't," agrees Mary.

"OK then," says John. "It's settled. Stephen, if you come with me up to the Deephaven office we'll arrange the swap."

"Great."

John returns in about half an hour and we begin moving all our stuff to the other cabin. I start counting the number of times I have carried suitcases today. From our home to the car – from the car to our airport – from the Boston airport to the rental car van – from the rental car van to the minivan – from the minivan to the truck – from the truck to the cabin – and now, from the first cabin to the second.

"Some vacation this is turning out to be," I think to myself.

Mark and I start carrying everyone's suitcases outside to Matt and John who are waiting at the front door because they are still too skittish to come back into the cabin.

"Just drop everything here and we will take them over." says Matt.

"Fine."

I leave Nan's big suitcase until last, and after I roll it to the top of the stairs, fatigue and frustration take over, and I give the obese bag a hard kick, and watch as it careens down the stairs. For some reason, that makes me feel better.

"What happened?" asks Mark.

"Oh, sorry. It just slipped out of my hands."

It takes less than a half hour to get everything moved. Not as big a job as I had thought. The new cabin has a different layout than our old one, but is almost identical in all other aspects. Same type of fireplace. Same pull chain lights. Same windows with the old wavy glass that distorts your view outside. Same creaky floorboards. Same white birch railings on the stairs. Same rocking chairs. I guess it will be fine.

Nan and I take some time to unpack. All the bedrooms have large old painted dressers with various scratches and chips, and drawers that sometimes stick. The closets are built out from the bedroom walls, usually in one of the corners, and have wooden doors that latch closed with hooks or small rotating wood blocks.

The walls throughout the cabin have knotty pine paneling on one side and open, rough-cut 2 by 4 studs on the other – which essentially means no soundproofing. You can hear just about anything, anywhere in the house, which I presume, is why the twin beds have never been replaced with something more conducive to sleeping together. I remember as a kid, waking up in the middle of the night whenever anyone got up to use the bathroom. First you hear the bed springs, then the floorboards, then the bathroom door closing, and finally, the flush. And then, once more when returning to bed, the door, the floorboards and the springs. I guess that's why they call this place a family camp.

"Nan, which drawers should I use for my clothes?" I ask.

"Oh, sorry Luke. I think I used them all."

"Are there any hangers left in the closet?"

"I think so, but you know, it really doesn't matter, the closet is full. Extra hangers aren't going to help much."

"So, where do you suggest I put all my clothes?"

"Oh, you can just live out of your suitcase. You didn't bring many clothes anyway."

"Fine. Would you like me to lock up any of your jewelry in the safe?"

"Oh, I didn't see it. Where's the safe."

"It was a joke, Nan."

"Oh."

About that time we hear the dinner bell ringing. The bell is in a small tower outside of the dining hall and the first kids that line up in front of the tower before each meal get to pull the rope and ring the bell – a simple activity that produces an extraordinary amount of excitement for all the kids in the camp.

There has always been something Palovian about the bell because anytime my brothers and I would hear it, our conditioned response would be to drop everything we were doing and run to the dining hall – whether we were hungry or not. Today is no exception, and we all routinely file out of the cabin and walk briskly along the lake to dinner. About a five minute trek.

The food at Deephaven is sometimes great and sometimes mediocre. But it is always plentiful and it is always extremely difficult not to overindulge – especially at the first meal. Everything is served buffet style, and there is a tendency to take more than you can eat. So, after a bowl of homemade soup, a salad heaped with fresh vegetables from the salad bar, a bowl of spaghetti and meatballs, roast beef, mashed potatoes and gravy, carrots, string beans and a large piece of the 'welcome back' cake with chocolate ice cream, I am ready to waddle back to the cabin. So is everyone else.

On the way back we all vow to cut back on our portion sizes at future meals.

Fat chance.

We return to the cabin in time to see the sunset, and we move all the rocking chairs from the cabin onto the porch – to watch the orange sky roll up and down as our chairs rock back and forth.

After a few minutes, Mark opens the bar and begins preparing drinks for everyone. Nan and I start out by sipping a wonderful French Bordeaux, and Matt and Deb indulge in a couple of 4 olive martinis. John has some type of fruity rum thing and Mark refills his scotch. Even Mary has a Chardonnay – a bit unusual for her.

We talk for a while – a series of dull but polite conversations. We discuss the upcoming wedding, our first dinner, the beautiful lake, the gorgeous sunset, the rustic cabin, and the weather. But by the time it is dark, it seems as if we don't have much more to say to each other.

It's interesting how five kids growing up in the same house with the same parents could have traveled such divergent paths. I'm sure that some of our differences were due to heredity – the way we were born, but I'm also sure that our many varied experiences were contributing factors.

Even when growing up, we had trouble getting along with each other. We all had such different personalities and interests, and squabbles and disagreements were always a part of our lives. And we each were affected differently by our strong-minded father who, on most occasions, was the commander-in-chief of the house. Dad had strict, unyielding standards for raising children, and he posted a weekly list of goals, activities and chores that were mandated for each of us, and if any of us veered off track, we were called to task.

We all were given the same expectations to meet – to excel in school, to help out around the house, and to be actively involved in church, Sunday school and church youth group every week. But all our childhood experiences affected us in different ways.

Matt excelled at school, but he wasn't interested in any church related activities. He was, however, smart enough to realize that he had to go through the motions to please Dad. And he did. He was the first born and always seemed to be Dad's favorite son.

I was the second child, and I don't think Dad ever liked me very much. I wasn't nearly as smart as my older brother, and I didn't excel in much of anything except baseball. But Dad believed education was more important than sports, and I often received the brunt of Dad's wrath when my grades were below par, or when I was caught skipping church activities for batting practice. I spent much of my adolescence in trouble and on restriction, until I realized that if I kept my mouth shut and did what was required, I could survive. And that's what I did.

Mark, the third in line, was always the pious one, and everyone in the family always knew he would eventually go into the ministry. His easy going nature, and propensity for reading the Bible kept him out of trouble most of the time. It wasn't until later that we learned that he would frequently sneak out of the house after curfew to rendezvous with a female friend – a foretelling of problems to come.

John, the youngest son, was sensitive and shy, and could always be found hanging around Mom's coat-tails. We always thought that he was probably gay, but he never let on to his sexual preferences, and Dad and Mom never suspected that he might be homosexual.

And Mary. Well, Mary was the spoiled baby. She was the only one in the family able to get Dad to change his mind or make him agree to compromises. She usually got her way, and over time both Mom and Dad lost control, and Mary became a rebel.

And so, here we are together – with nothing in common, and nothing to say. A room full of siblings, each with different personalities and experiences, who have, over time, grown farther and farther apart from each other. And now, with different lifestyles and vocations, the chasm between us has grown larger, and there doesn't seem to be much hope for any of us to salvage relationships that have deteriorated so far over so many years.

So, we just sit there, all of us staring at the walls wondering why we ever thought getting together would be a good idea.

But after a long awkward period of silence, John says, "Matt, do you believe in ghosts?"

That gets everyone's attention.

"I certainly don't," I blurt out before Matt has a chance to answer. I know where you are headed with this – and if you think our old cabin next door is haunted – then you're as crazy as all those ghost hunters on TV."

"I was asking Matt"

"Sorry."

"Well," says Matt, "I'm not sure, but over the years I have come to believe that there might be some type of spiritual world out there ... someplace."

"I don't know how you can say that," I respond. "When there is absolutely no credible scientific evidence to support such a notion?"

"I know."

"So?"

"Well, you're right, there's no direct evidence. And you know, I used to believe, just as you do, that ghosts were just figments of the imagination."

"So what changed your mind?"

"It's my fault," says Deb as she waves her hand in the corner. "He's been married to me too long."

"That's true. I mean it's true that it's Deb's fault. It's not true that we have been married too long. I love you darling."

"That's sweet of you to say, but you're still going to have to sleep in your own twin bed tonight."

"I know."

"So, what does Deb have to do with this?" I ask.

"Well," replies Matt, "I'm sure you have all noticed that Deb, at times, has some odd behaviors."

"Like this afternoon on the porch?"

"Yeah, and early in our marriage, I used to think that these behaviors were some type of psychosis. Deb herself, knew something was wrong, and we spent a number of years trying to find various ways to treat, what we considered, a physical, or perhaps a mental problem."

"You always thought I was crazy," says Deb.

"I did for some time. So did a lot of other people. It was the main reason all the adoption agencies turned us down after we discovered we could not have children."

"So what was the problem?" asks Nan.

Matt looks over to Deb, and Deb nods her head in approval. And Matt continues, "Well, you see, it's difficult to explain – mainly because I don't understand it. But ... somehow ... at unspecified times ... through some unexplained mechanism ... Deb has some type of connection ... to another place in space and time. You might want to call it a spiritual world ... or maybe something else ... I don't really know ... but there is some type of connection that cannot be explained with our current scientific understanding of the way the world works."

"I don't believe we are having this conversation," I reply. "No rational person believes in spooks. Spirits only exist in cartoons and in Hollywood movies."

"What about you Mary?" asks Matt, ignoring me for the moment. "Do you believe in ghosts?"

"No, not really."

"But Mary, you believe in angels. Don't you?" asks Mark.

"Of course. Angels are in the Bible, and I believe everything in the Bible."

"Well," responds Mark, "if you believe everything in the Bible, you not only believe in angels, but you also believe in ghosts. There are evil spirits all throughout the New Testament. And the Bible not only talks about evil spirits, but also about demon possession and exorcism."

"I guess I never really thought of ghosts and spirits as being the same thing."

Mary is the type of person I refer to as a Christian novice – a recent convert who hasn't completely worked through all the details of her faith. Mary has been attending a fundamentalist Christian church in suburban Atlanta for a while now, and has been indoctrinated into many of their beliefs. But her faith is built primarily on a foundation of emotions and feelings – and she lacks a broad understanding of the underlying theology she professes. She's no match for Mark, who is an ordained minister with many years of formal Biblical training and a doctorate of divinity degree from Duke University.

Mary is 10 years younger than my youngest brother, John. It is the consensus of the family that she was a surprise baby, although Mom and Dad told us that they always had plans to have a baby girl. Mary was a junior in high school when Mom and Dad had their car accident, and she had more difficulty than the rest of us dealing with their deaths. As an adolescent, she was always a little wild and uncontrollable, but after the funeral, she began a downhill spiral. She became involved with a bad group of friends who got her involved with various drugs, and when she turned 18, and was able to leave her guardians, she ran away to Florida with a 30 year old guy she met at an amusement park.

Over the years, Mary has gone through almost a dozen bad relationships, the last one breaking up when she was six months pregnant.

But a baby turned out to be a good thing for Mary – a life changing event. From the moment she first held that little boy in her arms she vowed to turn her life around. She eventually moved to Atlanta and began working as a waitress in a high end restaurant in the suburbs. About the same time, she started attending an independent fundamentalist church close to her apartment. It wasn't long after that, that she was saved – born again. And now, she is an adamant follower, who hangs on to every word that Pastor David proclaims from his pulpit. If Pastor David says something, it is as if God himself is speaking.

Mark, on the other hand, has followed a different path. After he graduated from Duke Theological Seminary he moved to Texas and was ordained as a Methodist minister. Over the following 16 years he rose quickly through various positions in the church, eventually becoming the senior pastor of a large congregation in Dallas. Mark was a well-respected member of the community, hosting local events and giving speeches and presentations at various area functions – until one Sunday morning, when a local newspaper published a story accusing him of having sexual relations with over eight women in his congregation. Mark tried to deny the accusations, but was eventually forced out of the church, and a few months later he had to file for bankruptcy because of civil lawsuit judgments against him.

During this difficult time, Mark became bitter and completely lost his faith in the church, and some of his faith in God.

Mark's weakness for beautiful women has been his ultimate downfall, and since leaving the ministry, he has spent a number of years trying to live a middle-aged swinger's lifestyle – where a man in his forty's dresses and acts like someone twenty years younger. It's a difficult thing to do, but Mark has managed the role relatively well, even with a receding hairline and few excess pounds around his waist. He has a certain charm and openness that is inviting, and his good natured approach and sense of humor have broken down the defenses of countless eligible females.

Mark has made a decent living over the last 12 years, by selling life insurance, mostly to former female church members. He has had numerous girlfriends over that time, but never seems to be able to hold on to one person for very long. He just recently ended a two year long relationship – one of his longer ones.

"You know," I say, "I really do not understand how anyone can believe all those Bible stories about evil spirits and casting out demons. It just makes no sense to me at all."

"Well," replies Mary, "if you deny what is written in the Bible, you're going to end up in Hell."

"Mary, I don't think that Luke is going to Hell," says Mark.

"You're gonna find out soon enough, Mark," responds Mary, "because that's probably where you are headed as well."

"Mary, you don't know what you're talking about."

"I do too."

"Hold on just a minute," says Matt. "Let's approach the subject of spirits and spiritual places from a scientific perspective."

"Is that even possible?" I ask.

"It is in my opinion. But I'll let you decide. Mark?"

"Yeah"

"Do you still have the box that the scotch came in?"

"Sure. Here it is."

Matt walks over and gets the box from Mark and says, "OK, everyone. You have to understand. Things are not always as they appear, and the latest advances in Physics and Quantum Mechanics demonstrate to us that the world we live in is a much stranger place than we think. Now, let me ask you a question. What is inside this box?"

"Well," says Nan, "there was scotch in the box, but I think Mark has already finished off most of that."

"So what's in the box now?"

"Nothing," says John.

"That's not true. There's nitrogen in there. And oxygen. And carbon dioxide – even though you can't see them."

"I know where you are going with this argument," I interrupt, "but there's a big difference between invisible gases and invisible ghosts. Even though I can't see what is in the box, I can use instruments to determine what's in there. And after I finish my testing, I can tell you precisely the percentage of each gas the box contains. I can even tell you the number of molecules. You can't perform the same type of analysis on this fictitious spiritual world of yours. You can't tell me the number of molecules it has – or anything about its composition. We just invent the nebulous concept of a spiritual world in our heads to explain things we don't understand."

"Hang on. We're not done yet. Let's pursue this a little further. Let's say we take this scotch box into outer space. Now what's in the box?"

"Nothing," replies John a second time.

"That's still not entirely true. There are still a couple of hydrogen atoms hanging around, even in deep space. So we will take a small pair of tweezers and pick those out. Now, John, what's in the box?"

"I'll try this one more time. Ah ... nothing."

“No!” snaps Matt. And after a short dramatic pause, he says, “There is space in the box. And space is a something. It has a structure, and we call this structure the ‘fabric’ of space.”

“But there's nothing there. It's just an empty vacuum.” says John.

“I know that's what most people think, but the fact is that there is something there. Even though you can't see it, there is space in there, and space has a framework – a structure.”

“That's just hard to imagine.”

“Let’s approach this a different way. We know that the universe is expanding. So, what is expanding?”

“I'm not going to answer any more questions,” says John. “I can't seem to get anything right.”

“I'll try this one”, says Nan, “The universe is expanding at the edges, constantly growing larger.”

“Not true. We don't even know if the universe has an edge, and if it does, we certainly can't see it. The only thing that we know is expanding is the empty space between things.”

“And how do you know that?”

“Well, if we look at a neighboring galaxy close to us, we can determine the speed it is moving away from us. Then, if we look at a galaxy twice as far away, and measure its speed, we find it is moving away twice as fast as the first. That's because there is twice as much space between us and the second galaxy, and all the space between us is expanding. A galaxy four times the distance from us has four times the space and moves away four times as fast. Six times the distance, six times the space, six times the speed.”

“I still don't understand how nothing can expand,” says Nan.

“It's not nothing. It's something. Space has a structure,” says Matt, “and it expands and bends.”

“It's just difficult to visualize,” says Nan.

“Well Nan, you may understand it a little better if we talk about gravity. Do you know what gravity is?”

“Sure, it's the force of attraction between two masses.”

“No.”

“You see Nan,” says John, “you are answering more questions wrong than I did.”

“Sorry,” says Matt. “I don't mean to be critical. Most people don't understand this, but the Newtonian theories of gravity were replaced in the early 20th century by Albert.”

“Einstein again?”

“Yup. You see, gravity results in two masses being attracted to each other, but that is not what gravity is. Gravity is the bending of space. Supposedly empty space. The same space that's inside our scotch box.”

“But how can empty space can bend?”

“You just have to accept the fact that space is not a void. It has a microscopic physical structure. And that structure – that space – bends.”

“OK, let's say we can accept that. It still doesn't explain how simply bending space can create gravity.”

“Well, it's really not that difficult to understand. You see, everything likes to travel in a straight line through the structure of space. Just as if it were traveling through a long straight tube. If you hit a golf ball in outer space where there is no air and insignificant gravity – that ball will always follow a straight line.”

“Even in outer space, I don't think my golf ball would follow a straight line,” interrupts Mark.

“Actually, it would. Even with a wicked slice it would still go straight. That is, until the straight line structure of space is changed. You see, if our golf ball encounters a large mass that bends the space around it, a mass like the earth – our ball will follow a curved trajectory, because the empty space around the earth is curved. It's just like the hills and valleys on a putting green. You hit the ball straight, but the uneven terrain of the green causes the ball to curve."

"So, I can blame all my bogies on gravity?"

"Well, partly, but the point here is that if you put any mass next to an empty piece of outer space – that space bends. And in order to bend, empty space has to have a structure."

"Can we talk about something else?" asks Deb. "I'm bored."

"Just one more important point," says Matt. "If we look at supposedly empty space on smaller and smaller scales, smaller than any instrumentation allows us, smaller than we can even imagine, we find lots of activity. It's not the quiet empty vacuum that most people envision. There's lots of activity going on at a microscopic level. We call it quantum foam. And in this foam there are all kinds of particles continually appearing and moving and then disappearing. And this goes on continually throughout the entire spacial fabric. The only reason we are not able to see this activity macroscopically, is because all these interactions tend to cancel each other out. An electron, for instance, can suddenly appear out of nowhere, and then a brief moment later, be annihilated into oblivion by its antiparticle partner – a positron."

"How can something come out of nowhere?"

"That's probably a topic for another discussion, but let me just say this. One of the newer theories that tries to explain the structure of empty space, and the behavior of small particles ... is called string theory. And according to string theory, when you look at the structure of very small pieces of empty space, you find something extraordinary. Wrapped up within each tiny piece of space there are ... six extra dimensions. We are all familiar with our three dimensional world – the one with height, width and depth, but according to string theory, there are six additional dimensions that we cannot see because they are hidden – tightly wrapped up within the very small pieces of the three dimensional space that we live in. But even though they are invisible, these six dimensions are all around us. They permeate all of space – throughout the entire universe."

At that moment, we hear a scream that seems to be coming from the cabin next door. Our old cabin. Then, another scream, longer and louder than the first. And then another.

I immediately jump up and run toward the porch door. I expect everyone else to follow, but for some reason, they are all frozen in their chairs. I suppose that everyone is scared to move because the screams seem to be coming from the cabin that Deb refused to go into earlier today. But that's just silly.

But it doesn't matter. I swing open the back door, jump down the steps and run as fast as I can on the dark, narrow, rocky trail leading to the cabin next door. As I round the bend, I see three teenage girls, obviously shaken, standing about 30 feet from the back door.

"What's wrong?" I ask while gasping to catch my breath.

And in unison, all three girls say, "Bats!"

"Bats?"

"Yeah," says one, "all over the house."

"It's OK, they won't hurt you," I say as I walk toward the back door. Through the screened-in porch I can see two teenage boys wearing bicycle helmets, swinging tennis rackets wildly through the air, trying to swat several small mammals flying by their heads. "It's OK," I repeat, yelling louder than I probably should have. "They won't hurt you."

One of the boys lets me in and says, "These suckers are fast! We can't swat them." Then he ducks as a bat flies past his helmet.

"It's OK. It's OK. All you have to do is open the door, and they will fly out."

"Really?"

"Yeah."

So, we open the doors and watch as four bats make a beeline out of the house. We then search the rest of the house, and find one more bat cowering in the corner of one of the rafters in the hallway.

"Do you have a towel?" I ask.

"Sure"

I grab the towel, climb on a chair, and gently pick up the poor little fellow, and then I release him through an open window.

"Is the coast clear?"

I turn around and see the teenage girls, and Stephen and his wife standing in the doorway. "Yeah, it's fine," I say. "These bats won't hurt you."

"But they were flying right at our heads."

"Yeah, but they will veer off, and go around you. They're pretty good at that. You can talk to the Deephaven office, and they will make sure that there are no holes in the cabin where the bats can get in. We had the same problem with bats the last time we were here, and the maintenance guys took care of it for us. But you have to understand, the bats are the permanent residents. You are just visitors."

"Thanks for your help."

"You're welcome."

As I begin to walk back to our cabin I see the rest of our cowardly clan on the path, huddled together, trying to catch a glimpse of the action, without having to get too close.

"Everything's fine," I say. "They just had a few bats." Then, we all walk back to the cabin and Mark makes everyone another round of drinks.

The night has cooled off considerably, so we move the rockers back inside and start a fire in the fireplace. I think everyone was feeling a little chilled, both physically and emotionally, and the fire seems to help.

"Remember how we had bats in that same cabin the last time we were here?" asks John.

"Oh yeah, I remember," says Mark. "We tried to chase them out of the house with our fishing poles."

"And you poked me in the back, as I remember," says Matt.

"And Mom and Dad slept through most of it."

"I don't know how they did that. We were screaming almost as much as those teenage girls."

"So Deb," says Nan. "Do you think that maybe the reason you were afraid to go into that cabin ... was because somehow you sensed that the bats were in there?"

But before she has a chance to answer, I interrupt, "You know, that could be a plausible explanation. People do sense things in different ways – ways that are sometimes difficult to understand. Like ESP. But just because we can't fully explain something doesn't imply that there is some type of supernatural intervention. It just means that our science has not yet advanced to the point where we understand every phenomenon we encounter."

No one says anything for a moment, so I continue – trying to reiterate my point, "You know, perhaps Deb subconsciously heard the bats, or somehow felt some other type of vibration they were emitting. That is entirely possible. But you have to understand, if Deb did pick up some type of signal from the bats, it was through a process that we can study – and eventually, understand."

"So Deb," says Nan, "do you think you sensed the bats? Is that why you had so much trouble going into the cabin?"

And without a moment's hesitation, Deb quietly and deliberately says, "No, it wasn't the bats."

The fire is now roaring, and we all find ourselves staring into the flames – watching the yellows and oranges dance across the logs in an ever changing pattern. It's hypnotizing.

"Matt." says John without looking away from the fire.

"Yeah."

"You know how you said that empty space has a structure?"

"Yeah."

"And that it stretches and bends?"

"Yeah."

"And that things appear and disappear within this empty space structure thing?"

"Yeah."

"And that if you look at a little tiny piece of space, it has other dimensions?"

"Yeah, another six dimensions in addition to the three we know of. Actually it may be seven. You have to understand, however, that it is still a theory. It is what many physicists believe, but it is still unproven."

"OK, then," says John, as he turns his head toward Matt and away from the fire, "let me ask you a question."

"Alright."

"Is it possible that somehow there's some type of spiritual world lurking in all those extra dimensions? Could that be where heaven is?"

"Well, that's where I was headed before we were interrupted by the bats. Extra dimensions in space open up an entire world of fascinating possibilities – from spiritual realms to parallel universes – stuff you once thought was only possible in science fiction novels."

"Really."

"Yeah, but you have to be careful. Anything we conclude about extra dimensions would be pure speculation. So, until you develop the tools that would be required to probe into those dimensions, and until you have some hard scientific evidence that they exist throughout all of space, you really can't draw any definite conclusions. You can speculate, but speculation is not good science,"

"Now, finally, that's something I agree with," I say.

"I never really disagreed with anything you have said Luke, I am just a little more open to the possibility that science is going to reveal to us things that we don't expect. This is an exciting time in physics because strange new things are being discovered every day. And the more we discover, the more we realize that things are not always as they appear, and there are even some people that believe that our macro perception of our world is, to a large degree, illusionary."

"You know it's funny," says John, "but remember how Mom and Dad always used to make us go to Sunday school?"

"Oh yeah," I reply. "And remember how we all hated it, and how we would try to act sick so we wouldn't have to go."

"Yeah, and remember how Mrs. Carpenter used to bribe us with candy to try to make us memorize all those Bible verses?"

"Yeah."

"Well," says John staring back into the fire, "I have long forgotten most of what Mrs. Carpenter taught us. But I do remember one verse ... I guess because I thought it was such a strange quotation, and I never really understood what it was talking about. Anyway, the verse says something like 'heaven can't be observed, because the kingdom of God is among us', and I have always wondered how heaven could be all around us, but not be seen."

"And in the Bible," adds Mark, "both John the Baptist and Jesus say that the kingdom of heaven is 'at hand'."

"Yeah," continues John. So, I was thinking, maybe heaven really is among us ... or at hand ... but we just can't see it ... because it's hidden ... wrapped up in all those extra dimensions."

"It's possible," says Matt, "but again, it's just speculation, and there's no way you can approach the subject scientifically with our current technology."

"And so," I say, "when you make speculations like that – when you try to turn a discussion about empty space into an explanation of heaven based on a few obscure verses in the Bible – you are not dealing with science any longer, you are dealing with philosophy. And philosophy has led man astray countless times throughout his history. Matt just said that it was dangerous to speculate on this, and I agree. It just leads to erroneous misinterpretations. Just look at what has happened in the past. Whenever the ancients didn't understand something they made stuff up. They speculated, they perpetuated myths and they invented gods. But we don't do that any longer. Now we investigate and experiment and probe, and we don't draw any conclusions until we are able to explain things rationally."

"That's true Luke," says Matt, "but in the field of physics, the more we experiment, the more we probe, and the more we learn ... the more we realize that the world we live in is a very strange place."

"I'm tired," I say. Actually, I'm just tired of listening to my big brother, but I stop short of saying that. "I'm going to bed."

"Me too," says Deb. "Good night everyone."

"I think we will turn in as well."

"Me too. 'night."

"Good night."

"Good night all."

"Good night."

"I might just stay up a little longer and watch the fire," says John without looking up. "I'll turn in shortly."

CHAPTER 3 – DAY 2

Gravitation is not responsible for people falling in love.
Albert Einstein

Ice Cold Morning

"Luke, Luke. Are you awake? Luke?" whispers Nan.

"I am now," I respond after I roll over.

"It's morning, and you need to get up."

"Why?"

"Because it's freezing! And somebody needs to make a fire."

"I'm not getting out of bed. It's too cold."

"Somebody has to."

"Yeah, but there are six other people in this house. Including you."

"Luke!"

"OK, OK. I'm getting up."

Even though it is technically early summer, the last week of June can be pretty cold in northern New Hampshire – especially on clear nights when there are no clouds to hold in the heat from the afternoon sun. The fireplace will keep the cabin warm for a few hours after everyone turns in for the evening, but because there's no insulation, by morning the temperature inside is about the same as it is outside.

I muster enough courage to throw back the blankets and sit up on my slightly-too-short twin bed, and then attempt to slip on a pair of pants, a shirt and a jacket before I begin shivering. I look over to Nan, to see if she appreciates my sacrifice, but it appears as if she has gone back to sleep. "Oh well."

I sneak down the stairs, trying not to wake anyone, and throw some kindling into the fireplace along with the sports page of yesterday's newspaper. On the mantle is a box of safety matches, and as I strike one of the wooden sticks against the side of the box, I think to myself, "It's been years since I have used real matches. I'm not sure we even have any left in our house. All we ever use now are those clicker lighter things."

After the kindling flares up, I throw on a couple of larger logs and try to warm myself by rubbing my arms together, as I wait for the fire to heat up.

Before too long I begin hearing bedsprings and floorboards throughout the cabin, and one by one, everyone conjugates in front of the fireplace. No one says much – probably because we all are in need of a warm caffeine fix to thaw the frozen blood in our veins. Nan, the last one down, arrives just about the time the breakfast bell rings, and everyone immediately abandons the fire and marches off to the dining hall. Time to eat.

After breakfast we walk back to the cabin on a different path, through the woods instead of along the lake. Not far from our cabin we pass the ice house, where huge blocks of ice are stored for the all the cabins' ice boxes. We stop for a minute to take a look. Inside, we see a strong good-looking young man moving one of the 130 pound ice blocks out of the sawdust with a two prong metal hook – a tool most people have only seen in movies that portray life in the early 1900's. He pulls the block outside onto the wood decking, washes the sawdust off with a garden hose, and loads it onto an old green wheelbarrow – another remnant of an earlier time. The wheelbarrow has a flat wood bottom with removable wood panels on the sides. There's a balloon tire in front and two wobbly wooden legs supporting the rear.

"Good morning. My name is Peter."

"Morning."

"You guys in the cabin down this path?"

"Yeah."

"Great. I'm headed over there now," he says. "I'm running a little late this morning. I had to check your neighbor's cabin for bats."

"Find any?"

"No, but we found a small hole in the rafters that we covered up. That took some time, and now I have to catch up on my ice duties."

The process of delivering ice to the cabins hasn't changed a bit since the last time we were here. For that matter, it really hasn't changed any for over a hundred years – from a time before Kelvinator, Frigidaire and GE began mass producing new-fangled electric refrigerators for home use.

"Do you still cut the ice out of the lake?" asks Mark.

"Sure do," replies the young man. "Well, actually I don't. It's way too cold for me. But a rugged crew of old timers come out to the lake every January and fills our ice house with about 2000 ice blocks. Something they've done for years. And you know, we may be the last place in New England that still continues this tradition."

"You still use horses to help cut the ice?"

"Oh no. We now use a 38 inch, gas powered circular saw, but we still have some of the old saw blades the horses used to pull across the ice."

"How do you get the ice out of the lake?" asks Mary.

"Well," says Peter as he parks his wheelbarrow in front of our cabin, "we wait until the ice on the lake is between 11 and 15 inches thick. Then we plow off the snow and cut out 16 by 20 inch ice blocks with our saw. The blocks are then floated toward the shore in a channel we create in the ice. Then we load the blocks onto a wagon and transfer them to the ice house."

Our ice man pulls a pick out from his back pocket and chops off a hunk of ice for our cabin. I hold the screen door open for him as he picks up the ice with those old metal thongs and drops the block into our icebox just inside.

"How do you fit 2000 blocks of ice inside that small little ice house."

"Well, it's a bit deceiving. Much of the ice is underground. You are only seeing the blocks on the top."

"And it doesn't melt?"

"Some, but we pack the ice blocks close together so air can't get in, and then we surround everything with sawdust. It's great insulation. And the ice lasts well into the fall."

"You know," I say, "it seems like a lot of work. Wouldn't it be a lot cheaper just to buy refrigerators for the cabins?"

"Yeah, it would. But it wouldn't be the same. Would it? I mean, you might as well stay in a hotel. Most of them have refrigerators. But there are some things you will only find at Deephaven. An ice box is one of them. It's funny, you know, when you go back to doing things the way they did things 100 years ago ... there seems to be less stress. Things are not as complicated – and it's easier to relax."

"Maybe for us," says Deb. "But Peter, you're the one that has to carry around 130 pound blocks of ice."

"Yeah. I guess so," replies Peter. "But you know, I love working here. I feel part of all this. I have a connection here. A connection with nature. A connection with all the families that come back year after year. A connection with this old camp. My great grandfather was one of the first people to build a cabin at Deephaven. Back in 1907."

"Wow."

"Well, see ya tomorrow."

"Thanks, Peter."

The squeaky spring pulls the screen door shut, and as we walk out toward the porch, Deb says, "You know, I have a good feeling about Peter."

"OK guys," I say. "Let's go. It's tennis time. I have been looking forward to this match for a long time."

"I'm coming," says Matt. "You still playing with that old wood racquet?"

"No, but if I were, I would still beat you."

It has been over 40 years since Matt, Mark, John and I have played tennis together. The last time we were here. Actually Matt, Mark and I were the only ones to play. John was only about 8 years old, and was relegated to chasing the balls we hit over the fence.

"Nan, you and the girls coming with us?"

"I don't think so Luke. There's a lecture about the geology of this area in the playhouse, in about half an hour. And I think Deb, Mary and I will drop in after we stop by the camp store."

"OK see you after the game."

Deephaven has eight clay tennis courts that are located at four different locations around the camp. I signed up for one of the more remote locations, because I wasn't sure we were ready to play in front of a large public audience. As it turns out, it's worse than I had imagined.

We walk out onto the court to begin warming up, but after Mark hits two or three balls he abandons his position and walks over to the adjacent court – and begins a conversation with the two cute women playing next to us. That leaves John on one side by himself, and, as we learn after a few minutes, John is not a very good tennis player. He has trouble returning every ball – as he prances uncontrolled from one side of the court to the other – swinging wildly at our intentionally slow serves, with a wobbly wrist and very little back swing. It's a little embarrassing. I'm just glad no one else is watching.

"Keep your eye on the ball, John."

"I am."

No one in the family really knows much about John – his tennis game or anything else about him. None of us have seen him in years.

Midway through his senior year of high school, John made an announcement to our parents about his sexual orientation. But his attempt to come out-of-the-closet did not sit well with Dad. The phrase 'alternative lifestyle' was not part of Dad's vocabulary, and Dad's way of dealing with John's homosexuality was to do everything in his power to change John's sexual preferences. So, John had to endure over six months of grounding and restriction, over six months of screaming and yelling, and over six months of forced therapy at a conservative Christian counseling group that believes that all homosexuals are going to Hell.

And after over six months of Hell on earth, John left home – for good. The day after he graduated from high school, he bought a car with the money he had saved – and drove to San Francisco.

And, no one heard anything from John for about ten or fifteen years.

But eventually, John began calling on holidays and on our birthdays, if he remembered, and gradually, over time, we have begun to keep in better contact.

John has done relatively well in California. He has opened up a chain of flower stores with his friend and partner – who he has lived with for about twenty years. We often kid him, saying that the flower business is a bit of a typecast, but he just laughs and says that floral arrangement was just a natural talent he was born with.

Apparently, tennis is not. "John, get your arm back! And step into your swing! And keep your eye on the ball!"

"I'm trying."

John still exhibits many of the emotional scars from his childhood, and he is often withdrawn and aloof. He has always been very sensitive, a little shy and a bit insecure, but growing up with an intolerant and overbearing father, and a complete lack of any support, financially and emotionally, during his early adult years has had a pronounced effect on his personality.

"John. How many times do I have to tell you? Keep your eye on the ball!" I scream.

"You know," says John quietly, "maybe I'll just watch the rest of the game. I really don't feel like playing."

"I may have to take a rain check too," says Mark from the court next to ours. We look over and see that Mark has apparently made friends with the two women playing there. "We are going over to look at some of the old Deephaven artifacts and books they have displayed over near the dining hall. Oh, by the way, this is Dawn and this is Hope. They have a cabin three or four down from ours."

"Great. Some game this is turning out to be. So much for a reunion match," I mutter under my breath. "OK, Matt, you and I can play,"

"Nice to meet you two," says Matt.

"You too."

John takes a seat on the side of the court and Mark begins to pack up his equipment – and Matt and I start volleying. Back and forth, forehand to forehand to backhand to backhand. Eight, ten, twelve times over the net – a great volley for not having played for so long. It feels good, and smashing the ball is a good release for some of my anger. I had thought that this game would be a bonding experience for my brothers and me – something that would remind us of a time long ago when we enjoyed playing together – and being together. But today, that is not to be.

The clay court slows the ball down considerably, and makes it easier to see, and easier to hit than on a hard surface, which may be why my backhand swing is working so well today.

Twenty, twenty-two, twenty-four times back and forth across the net. What a great volley. It seems it could go on all afternoon, but then I seize an opportunity and pounce on a slow hit ball – and make a great shot down the line which I am sure is un-returnable. But Matt reacts quickly, and to my surprise, connects and sends the ball back to my side of the court. But he pitches the ball up high, close to the net, and I run up and slam it right toward Matt's feet. Matt is unable to retreat in time and the ball hits him hard on the calf, and he falls to the clay. John, Mark and Mark's two new friends rush out to see if he is alright. He appears to be fine, but with everyone paying attention to Matt, no one sees me, holding onto the net with both hands, trying to catch my breath. I think, for a moment that I am going to faint, but gradually, my strength and my balance begin to return.

"You OK?" asks Matt.

"Yeah, what about you?"

"I'm fine. But I think I have had enough tennis for the day."

"Me too."

On the way back to the cabin I stop by the Playhouse, to see if Nan is still there listening to the geology lecture. I walk in the back and spot Nan, Deb and Mary in the second row. There are only about a dozen people there, and they all appear to be falling asleep – and after listening for a few minutes, I can see why. The lecturer is an old guy with a grey beard and a monotone voice, and he is talking about how the mountains around Squam Lake were formed by plutons (whatever those are) between about 350 million years ago in the Devonian Age and 150 million years ago in the Jurassic. Then he starts talking about feldspar and tonalite and syenite rocks, a topic I am sure no one in the room cares anything about – and then he goes on to explain how, in the recent past (the last few hundred thousand years) the lakes and valleys around us were carved out of the softer rocks by a series of glaciers.

That is enough for me. I am starting to fall asleep standing up. So, I quietly sneak out the back door. I'm sure Nan would have followed me, but it's difficult to leave a lecture when there's a small audience and you are sitting in one of the front rows.

An Afternoon on the Lake

By mid-afternoon the cold morning breeze has morphed into a warmer southerly flow, and with relatively few clouds, the temperature has risen into the upper 70's. A perfect day for being on the lake.

One of the most enjoyable and memorable experiences at Deephaven is spending a warm summer day on the water, and the camp encourages water activities by maintaining a large rental inventory of things that float. I guess we should have coordinated our boat requests better, because somehow we ended up with a small armada of watercraft tied to our dock – two canoes, three kayaks, a water bike and a small sailboat. Way more than we needed, but I think we all had the feeling that spending time on the lake together would be therapeutic – just like when we were kids.

"That didn't work with tennis," I think to myself. "I doubt it's going to work with water sports. Besides, when was the last time any of us was in a kayak?"

We all change into shorts and bathing suits and congregate down on our dock. By the time Nan and I arrive, Matt has already taken out a kayak and has paddled over to the far shore across the cove. I guess he didn't want to wait for us.

Mary briefly sticks her foot in the water and instantly screams, "Ahhhh ... its soooo cold!"

About that time, we hear a giant splash, followed by a loud "Whooo-eeee! This feels great." We all turn to look, and see Stephen, from the next door cabin, bobbing up and down in the lake like a cork. Then we hear several more splashes and a few loud screams as the five other teenagers from the house join Stephen in the water.

"Come on in!" Stephen yells over to us. "It's not too bad after the initial plunge."

The water in Squam Lake is still pretty cold in late June. You are usually lucky if it warms up to 70 degrees, but right now it feels colder than that. Most people like to come to Deephaven in July and August – when temperatures are a bit warmer.

"Yeah," I yell over to Stephen, "but that initial plunge may kill you!"

"Maybe," says Stephen as he dunks his head beneath the water. "But what a way to go. This is beautiful!"

The five teenagers swim out to a floating platform, about 30 yards from Stephen's dock and begin to play king-of-the-hill – laughing uncontrollably as they push each other off the platform into the frigid water.

"Maybe," I think, "it's easier to get along when you are younger – before you develop an intolerance for values that are different than yours – before you have to face the harsh realities of broken relationships – before you realize that life is filled with injustice, and pain, and suffering, and death."

I begin to wonder if anyone in our family will ever get along. Maybe we are just too old – too set in our ways. It doesn't seem likely that one week together is going to change any of our attitudes or feelings toward each other. It might just make things worse.

Nan and I spend a few precarious moments getting into one of the canoes, trying our best not to tip over. It's not as easy as I remember. John, meanwhile, gets on the water bike and begins pedaling along the shore, while Deb and Mary sit on the edge of the dock.

"I think I will take out the sailboat," says Mark, "and sail over to Hope and Dawn's cabin – to see what they are doing," says Mark.

"OK," I say, as we all go off in our own directions.

Nan and I paddle out to the middle of our cove and watch as Mark hoists the sailboat sail and sets off from the dock. The wind is blowing directly along the shore and Mark has to tack out toward us in order to make any headway up the lake. He passes close to us and waves. Then, as he turns back toward the shore, in front of our canoe, he reaches his hand out – and splashes me with some of the frigid lake water. I try to splash him back with my paddle, but in my haste, I don't take the wind into account, and the water flies back into Nan's face.

"Luke!"

"Sorry Nan."

"Let's get him!"

"OK!" I say as we both begin paddling erratically after Mark. But we are too slow, and as Mark begins pulling away from us he smiles at us beneath the sail and waves one more time.

"We will get him back when we ..." but before Nan is able to finish her sentence, an unexpected gust of wind hits, and Mark's boat flips over. Capsized.

"Ahhhh, it's cold!" shouts Mark.

We paddle over toward the disabled craft, and are joined shortly by Matt in his kayak and John on the water bike. We all surround Mark, laughing and making disparaging comments about his sailing ability, as he attempts to right the boat by sitting on the centerboard.

"I'll get you back! Every one of you!" he says, as he flips the boat upright and pulls himself out of the water.

"You have to catch us first," says Nan. "Race you back to the dock!"

And that starts a torrent of paddling, peddling, and jibing as five middle age adults try to outdo each other in an 80 yard sprint. John takes the early lead with his water bike, but Matt quickly comes up from behind and is about to overtake John when John turns his bike into the kayak and one of the bike's pontoons lifts the front of Matt's kayak out of the water. Matt almost turns over, but quickly regains control.

But both Matt and John have lost momentum and Mark has taken the lead in his sailboat. It looks as if he will win this ragtag race, but as he gets closer to the dock the wind fizzles and Nan and I catch up, with Matt and John close behind. It's a dead heat 15 yards from the dock. It could be anyone's race. A photo finish. It's too close to call as all of us abandon safe landing techniques and simultaneously plow into the side of our small wooden dock.

Fortunately, Mary and Deb who had been sitting on the dock, dangling their feet in the water, had the foresight to retreat to land seconds before the collision.

"Who won?" asks Mark.

"I'm not sure," I answer as I pull myself up from the bottom of the canoe back into my seat. "I think it was a tie."

"You guys are acting like a bunch of kids," says Mary from the shore.

"Yeah, that was fun," says Matt as he lifts himself out of the kayak. "Here Nan. Let me give you a hand out of that speedy canoe."

"Thanks."

"Here's a towel Mark," says John.

"Thanks, I'm freezing."

We spend the rest of the afternoon sitting together on the dock, reminiscing. We start by talking about our memories of Deephaven – canoeing, playing tennis, ringing the bell before meals. And then we talk at length about growing up together. Playing hide-and-seek in our house on Third Street – having snowball fights in the backyard – sneaking out after curfew to meet our friends. Memories of the good times together that we seemed to have forgotten or taken for granted.

About midway through a discussion of how much we loved Mom's spaghetti sauce, John says, "You know, with so many wonderful memories, why do you think our family fell apart? Why is it that, from the time we were teenagers, we have never been very close?"

"Well," responds Mark, "you were the one that left home when you were 18."

"Yeah, but that was after the three of you had already gone off to college. This all started way before that."

"I was so much younger than the rest of you," says Mary. "Ten years makes a big difference. I always felt as if I was an only child. You guys just ignored me."

"You know, I think Dad had a lot to do with it – us not being very close." says Matt. "He wasn't very affectionate, and he had such a dominating personality. And I think at times, we redirected the animosity we had for Dad, toward each other."

"I'm sure we all did that to some degree," I say. "But you know, there's a lot more to it than that. We're all so different. You wonder sometimes how we can all be from the same family. Matt was always so smart and scientific."

"I've learned over time, that I'm really not that smart."

"Mark was pious and religious," I continue. "That is, until that whole sexual scandal thing got in the way of a promising religious career."

"Those things happen," says Mark. "Sex is a petty powerful thing."

"Even for ministers?"

"Oh yeah."

"I was always into sports when I was younger," I say, "but no one else in the family was ever very interested in playing ball with me. And later in life, after the accident and my divorce, I became a hermit."

"I think, for a time, you didn't even have a phone number."

"No, not for over a year. And then there was baby Mary, who over time became a rebellious street smart adolescent."

"Until I found Jesus."

"Yeah, and John, well John was ..."

"I was gay. And I guess I still am."

"So you see, the cards were stacked against us. You put five people like us in a house together for any period of time and there are bound to be confrontations."

I stop for a moment, after realizing what I had just said. I had implied that the five of us, together this week in this cabin, were destined to repeat the same conflicts we had experienced in our childhood. How could I have said such a thing? Maybe no one will notice.

"Gosh," says Deb, "it's only an hour and a half before dinner, and we all have to get ready."

"And there's only one shower!" says Mary.

"I'm first," says Nan as she runs down the dock.

"Go ahead y'all, I've already had a bath in this lovely lake," says Mark.

A Puzzle

"Alright Nan," says Mary as she pounds on the shower room door. "Your time is up!"

"Just give me a minute more. I need to wash the soap out of my hair. By the way, do you have any extra towels out there?"

"Let me check," says Deb as she begins searching all the hall closets. Nothing here ... nothing here ... Oh, here are some." She hands Mary a stack of towels and then she disappears around the corner, apparently in search of more. Deb returns about a minute later and says, "Well, no more towels, but look at what I found."

"A brown paper bag?" says Mary.

"Yeah, but not just any brown paper bag. Look at what it says."

I walk over to Deb and look at the bag. On the side is handwritten with a purple marker, "Our Big Happy Family". "So, there's a big happy family in the bag?" I ask.

"No Luke. It's a puzzle."

"So we have to guess what's in the bag?"

"No, no, it's a real puzzle. A jigsaw puzzle. It looks like it's from one of those companies where you send in a photograph and they make a puzzle out of it. Let's put it together."

"But it doesn't look like there's a picture to go by."

"No, there's no picture, but that just makes it more fun. You don't know what you're going to get."

"You're probably going to get a headache," I say. "It'll be way too hard. There must be over a thousand puzzle pieces in there. It doesn't sound like much fun to me at all."

"I'm going to dump this thing out on the table."

About that time Nan comes out of the shower wrapped in a towel that is almost too small to completely cover everything private, and Mary rushes in to take Nan's place in the small cramped shower room.

"You have five minutes." says Matt.

"OK, OK, I'll hurry."

I expect Nan to run upstairs and change before her tiny towel falls off completely, but instead, she scampers over to Deb's table. "I heard you talking about the puzzle. I love these things."

"Great. Help me turn all these pieces over."

When Mary comes out of the shower a few minutes later, also wrapped in a towel, she says, "It's your turn Deb. Can I help with the puzzle?"

"Sure. Take my place here. I'll be back in a few minutes."

"I'm just going to go upstairs and take a bath in that old bathtub with the legs," I say as I begin climbing the stairs. But it doesn't appear anyone is listening. "That's OK," I think to myself. "I could use some time alone."

So I take a leisurely bath. And I shave and get dressed. And when I come back down the stairs, I see six people huddled around the puzzle, desperately searching for pieces to the border.

And no one is wearing anything but towels.

"Luke," says Mark. "Come and join our toga party!"

"But it's almost time for ..." but before I can finish, the dinner bell starts to ring. That starts everyone scrambling. Towels go flying and modesty is abandoned as everyone rushes to their rooms in a frenzied attempt to get to dinner before the blueberry pie is gone. If only I could get Nan to move this quickly at home.

In less than ten minutes, we are all marching out the door. Time to eat. Again.

“I've got to stop eating so much!” says Mary on the way back to the cabin after dinner.

“Me too,” says Deb. “But I did steal some chocolate cake to snack on tonight. I hid it in my purse.”

“Do you have enough for all of us?”

“That depends,” says Deb as she cracks a smile. “You gonna help me with my puzzle?”

“Sure.”

Once back at the cabin Mark opens the bar and begins taking drink orders. “You know,” he says, we are probably one of the few cabins here with a bar. Hope and Dawn said that most of the families around camp typically drink a lot more water and lemonade than alcohol. Of course, I told them that if they brought their water and lemonade over here – that I would fix it up for them.”

“You can fix something up for me,” says Deb. “I'm always better at finding puzzle pieces after a few drinks.”

“I'm working on it now.”

So, while Mark is mixing drinks, the three girls begin diligently working on the puzzle. The rest of us pull up rocking chairs around the table – but a little further back – far enough back to avoid a confrontation with the front line forces, but close enough to be in a position to provide backup if the main troops develop battle fatigue.

“Ah Ha! Found it!” says Nan.

“Gimme that. I think it fits over here,” says Mary.

“No! These are my pieces. I'm working on all the light green colors.”

“But I know where they fit.”

“So do I. Let me over there.”

“This is my spot.”

“Better keep the bar open Mark,” I say. “This could be a long evening.”

"Yup."

"So, Nan," I say, trying to change the subject, "how was that geology lecture today? I dropped in for a few minutes in the back, and it sounded pretty boring to me."

"I don't know if it was boring or not," replies Nan. I was asleep."

"That guy didn't even know what he was talking about," says Mary.

"How's that?" asks Matt.

"Well, you know ... he was one of those evolutionists. He believed the earth, and everything on it just evolved by itself over hundreds of millions of years – and that's not biblical."

"I believe in evolution," says Matt.

"Me too," says Mark.

"What have I done?" I think to myself. Just when I thought we were at a turning point – just when I thought that we were beginning to get along with each other, we jump headfirst into another controversial subject – a subject I know no one will agree on. I shouldn't have brought up the lecture. I should have said something about our afternoon on the lake – or just kept my mouth shut.

"So," says Matt, "since you believe everything that the Bible says, do you think the earth was created in seven days?"

"No, I don't think that many people still believe that. But that doesn't mean that the Bible isn't completely accurate. You know, the sun wasn't even created until the second or third day."

"Actually, it was the fourth," adds Mark.

"Whatever. The point is that the length of a day wasn't even defined until the sun was created," continues Mary. "So, it might have taken God a little longer than seven days."

“My parents are Orthodox Jews,” says Nan, “and even though they interpret the Torah very literally, even they accept the fact that the earth is billions of years old.”

“You know,” says Mark, “you have to be careful with all timeframes in the Bible. They can't be taken literally. When the Bible says 'soon' it could mean thousands of years. When it says 'day' it could mean an age, and when it says 'a thousand years' it could be an eternity. The Bible itself says that for God, a day is like a thousand years.”

“It may take us a thousand years to finish this puzzle, says Deb. “We have a long way to go.”

“But time stretches and shrinks – depending on how fast you are going. Right Matt?” says Nan.

“Yeah, or when there's a lot of gravity.”

“Got it!” shouts Deb”

“I don't get it,” says Nan. “What does gravity have to do with time?”

“Got another one. That's another two puzzle pieces,” says Deb. “Can we discuss something other than gravity? Anything. How about the weather?”

“What do you mean time stretches and shrinks?” asks John.

“Oh, sorry John. You weren't with us in the car yesterday,” says Nan. “On the way to the camp, Matt explained to us how the faster you travel, the slower time passes.”

“Really?”

“Yeah, Matt's friend Albert Einstein said so.”

“Matt's friend?”

“Yeah. So Matt,” says Nan, “what did Albert say? What does time have to do with gravity?”

“Well,” replies Matt, “Albert realized that time and space are all mixed up together. You can't really separate them. So, if a large mass causes space to bend ... it also bends time.”

“Wait a minute,” says Mark from behind the bar, “How can time and space be mixed together? And how can you bend time?”

“This just doesn't fit,” says Deb. “I was sure these pieces connected.”

“OK you guys,” says Nan. “First, let me catch you up on how time works when you travel really fast.”

Nan then begins a two minute recap of the various things we had discussed on our car trip yesterday. She explains how light always travels at the speed of light, and how nothing can go faster than the speed of light. And because of this speed limitation, time has to slow down when you are traveling at very fast speeds. She repeats the example of two rocket ships speeding toward each other, each traveling at three quarters the speed of light, and explains that since they can't approach each other any faster than the speed of light, time has to slow down to compensate for their closing speed.

Matt just nods his head in approval, obviously proud of his star pupil who has quickly picked up on a few very difficult concepts.

Nan goes on to recap about the empty space in the scotch box, and how space has a structure – a structure that may have six or seven additional dimensions. And how this empty space structure bends when near large masses – creating gravity.

“How did I do?” asks Nan.

“Very well,” replies Matt.

“OK, so, why is it that when in a strong gravitation field, time bends as well as space?”

“Well Nan, it's really quite easy to understand. But first, let me ask you a question. What's the difference between acceleration and gravity?”

“Ah, I'm not sure. They're two entirely different things. Your question doesn't make sense.”

"Well, you know, Albert said that there's really no difference between the two at all. Acceleration and gravity are actually two manifestations of the same thing."

"What?"

"Yeah, let me give you an example. Let's say Mark jumps out of a plane 10,000 feet up in the air and starts accelerating toward the ground. Except for a little bit of wind in his face, Mark is weightless. He has overcome the influence of gravity on him, simply by falling or accelerating toward the earth. He has the same sensation as he would have just floating in middle of outer space."

"That is until he smashes into the ground. And dies!" says Deb without looking up from the puzzle.

"True, but we'll ignore that for the moment. Let me give you another example. Suppose John is out in the middle of outer space in an elevator, and he pushes the 'UP' button. Now, let's say that this elevator is pretty fast and can accelerate at 32 feet per second squared – exactly the same as the acceleration of gravity on the earth. From John's point of view, what's the difference between standing on earth, and standing in an accelerating elevator in the middle of outer space?"

"Well, in the space elevator, John won't be able to breathe," says Deb. "So he'll die too! Now, can we get back to the puzzle?"

"OK, OK. So, let's say we will pump some oxygen into the elevator so that John can breathe. Now, what's the difference?"

"Nothing?" says John timidly, knowing that up to now, he hasn't answered any of Matt's questions correctly.

"That's right. Nothing."

"Finally, I get one right!"

"You see," continues Matt, "Albert realized that there is really no difference between gravity and acceleration. You can create the same effect as gravity through acceleration –

like in the elevator. And you can use acceleration to completely negate gravity's force – like when you jump out of a plane.

"And therefore, it follows that if time slows significantly when you are accelerating to high speeds, it also slows significantly when you are under the influence of a large gravitational field – because gravity is really just a manifestation of acceleration."

"Why are we even talking about this?" asks Deb.

"I think it was because we were talking about the earth being created in seven days, and then Mark said that the Bible gets confused with times," I say.

"I can see how that can happen," says Matt. "Clocks on Jupiter tick a little slower than clocks on the earth, because Jupiter is so massive. And clocks on the sun tick slower still. But the effect of gravity on time becomes extreme when you travel to the center of a black hole – where the gravitational force is so strong that not even light can escape. There, something truly amazing happens. Time essentially stops completely."

"That's it! It was so obvious. I don't know why I didn't see it before," says Deb. "I finally found that missing piece. Time to stop. I need a break. Anyone want chocolate cake?"

Deb pulls out a large plastic container from her oversized purse and we all descend on the chocolate cake like a flock of ravenous vultures that hadn't eaten for a week.

"Hold on. Hold on. Let me cut it first," says Deb.

Deb divides the cake into seven small pieces and serves us our portions on paper towels. Mark takes the opportunity to refill everyone's drink and since it is beginning to cool down, I start a fire in the fireplace.

There's something magical about fires. They draw people in. They bring people together. I don't completely understand why. Maybe it's just the warmth. I don't know, but in what seems to be a conditioned reflex, everyone pulls their chair around in front of the fireplace, and begins staring at the flames. And as the fire grows warmer and brighter, disagreements about evolution and biblical accuracy seem to fade away, and we begin to discuss the timelessness of our wonderful camp – where nothing has changed for over a hundred years. Where the same cabins have hosted countless families just like ours, year after year. And where the same fireplace has warmed generations of souls gathered together around its bonding flames.

There's also something magical about red wine and chocolate. "They go so well together," I think to myself as I sip a cabernet and savor the last few crumbs of the cake. "This could turn out to be a great evening." I take a deep breath and breathe in the smell of the fire and the familiar aroma of the old cabin, but before I am able to exhale, Mark reignites the smoldering controversy, and the evening is once again ablaze – the warmth, camaraderie and magic of the fireside being incinerated by a lifetime of divergent views and values.

"So Mary," says Mark as he hands her a plastic cup of chardonnay, "do you believe in evolution now?"

"Certainly not!" responds Mary. "Pastor David says that evolution is a hoax that denies God's marvelous creative powers. He says that it is ridiculous to believe that man is just a product of random mutations and natural selection."

When I hear Mary use the words "hoax" and "ridiculous" my blood pressure begins to rise. I can feel my face turning red, but I try to control myself. I don't want to say anything I will regret, and besides, Mary really doesn't understand what she is talking about. She is just parroting words she has heard from Pastor David.

"And there are so many gaps in the geological record," Mary continues, "I don't know how anyone can accept evolution as factual."

"That was enough. Self-control will have to wait for another day. I try to figure out a way to tactfully refute Mary's inept statements, but my emotions get the best of me and I shout, "That's just ignorant!"

Everyone turns to look at me, and since I obviously have everyone's attention, I continue. "Evolution happened. And is still happening – without the intervention of any deity or supernatural hocus-pocus. I can't believe we are having this conversation. Matt, help me out here."

"Well," says Matt, "many people believe that evolution was guided by a higher being, but all the scientific evidence we have gathered so far, shows no evidence of a master designer."

"And as for gaps in the geological record," I say, "those gaps are being filled. It is a very slow, difficult and tedious process because fossils are difficult to find, and when you do find them, most of the soft body parts are not preserved. But despite all these obstacles, our understanding of the evolutionary process is remarkably complete."

"But eventually Luke," says Mary, "you are going to discover that the only way to explain some of the remaining gaps in our understanding ... is to admit that God was the ultimate creator."

"Science will eventually fill in the remaining gaps – not God."

"You know, it's interesting," says Mark. "Many people believe in what I call, the 'god of the gaps' theory. Whenever there is a gap in our understanding of the world, we simply attribute our lack of knowledge to God's intervention. It's happened throughout history. For example, the ancients didn't understand why the sun moved across the sky and attributed its motion the gods.

That is, until man eventually developed a better understanding of the solar system, and had to throw out the idea of some deity pulling the sun all day long. And the same thing happens today. Not too long ago, anti-evolutionists argued that flying birds could not have possibly evolved without divine inspiration because of gaps in the geological record. But the argument became moot when paleontologists discovered a rare fossil of a transitional dinosaur with feathers. So you see, all too often a gap we associate with God's creative powers is ... well, in reality, just an incomplete understanding of the world we live in."

"There will always be gaps," says John.

"Yeah," says Mary. "Like the Cambrian Explosion."

"What?"

"The Cambrian Explosion," repeats Mary. "It was a very short period of time when, out of nowhere, all of life as we know it today was created."

"Actually," says Matt, "it wasn't all of life as we know it today, but you are right. At the beginning of the Cambrian, over a ridiculously short period of time (geologically speaking), about five million years – the ancestors of almost every living organism in our world suddenly appeared. That includes arthropods, which later evolved into insects and crabs, and echinoderms which became modern starfish and sea urchins – and chordates, which became vertebrates, and mammals, and eventually, man. Before the Cambrian, period about 540 million years ago, all that existed on earth was some bacteria, algae and some primitive sponges and worms. And then BOOM. Suddenly, between twenty to thirty-five of the world's forty phyla appeared."

"And Pastor David says that the Cambrian Explosion proves that evolution is wrong," says Mary, obviously impressed with her knowledge of history.

"You know," I say, "most of us know about the Cambrian Explosion. There has been a lot written about it. But you have to understand, it's just another gap that will eventually be filled as we develop a better understanding of earth's history."

"You really think so?" asks John.

"Absolutely. Have you heard about Stardust?"

"Is that the stuff fairies sprinkle on people?

"No."

"Is it a new TV series?"

"No, it's a satellite that was launched by NASA to study comets. And in 2004, Stardust passed through the gas and dust tail of the comet Wild-2 and captured samples that were later returned to earth in 2006."

"And?"

"Well, subsequent analysis of the dust found something remarkable. Glycine."

"What's glycine?"

"Glycine is an amino acid that is one of the building blocks of life."

"But they didn't actually find anything living, did they?" asks John.

"No, but one of the working theories about the Cambrian Explosion is that meteorite and comet impacts transported some of the early building blocks of life to earth."

"But then where did the comets get them?" asks John. "Maybe every time you fill in a gap ... you create another."

"Yeah," replies Mark, "but you see, if you believe in a 'god of the gaps', every time a gap is filled in ... your faith in God decreases. Maybe just a little at first. But as time goes on, and as more and more of God's gaps are replaced with scientific explanations, your faith becomes seriously compromised ... and eventually you end up like me ... someone who was once very religious, who has, over time, 'evolved' into someone who is somewhat confused about what he believes."

"This is a dumb conversation. You all are way off base. You're missing the most important point here."

We all look up and see Deb working on the puzzle. In the heat of the conversation, no one had noticed that she had gotten up from her rocking chair in front of the fireplace and snuck back to the puzzle table.

"You all are concentrating on the physical," she says without looking up, "when you should be focusing on the spiritual. I don't care whether we evolved from monkeys or even amoeba – or whether there was a grand designer or a bunch of random mutations. That's unimportant. You should be discussing things that are more pertinent. Like consciousness."

"Consciousness?" I ask.

"Yes, consciousness," says Deb. "Consciousness is the key. Oh, I found another one. This puzzle is starting to come together."

"I'm not even sure I know what consciousness is," says Mary.

"I think it's kinda hard to explain," responds Mark. "I know I'm conscious. But I have no idea about the rest of you. As far as I know y'all may just be a bunch of robots."

"That's a good observation," says Matt. "Consciousness is hard to define and even harder to study because it's so personal. Scientists can study your brain. They can electrically stimulate it, and they can give you drugs that alter the transmission of signals between nerves, and they can watch your brain on a CT scan or an MRI. But these studies only shed light on how the brain functions – they don't even begin to address what goes on in the mind – our ideas and memories and feelings."

"So what exactly is consciousness?" asks John.

"I'm not sure anyone really knows what it is. So, we just try to describe it. We say that it is an awareness of everything around you – and it is being aware that we are aware of all those things – and it's a mechanism for generating ideas, and perceptions and memories and emotions and feelings – and then bringing them together in an organized and logical way – and it's a means to ..."

"That's way too complicated," interrupts Deb. "Consciousness is just your spirit. It's that little guy inside you that gives you your feelings and emotions."

"I don't believe in that!" I say abruptly.

"Believe in what?"

"You know, where you can separate the body from the mind. Where you have a soul that is immortal – that can live apart from the body. That's just not the way it works."

"You don't believe in a soul?"

"No. It gets back to what we were talking about before. I believe in evolution. And consciousness is just the product of an evolutionary advanced brain. When a brain reaches a certain level of complexity, we become aware of our surroundings."

"So, you just think that everything we do, see and feel can be explained by a bunch of randomly firing nerve bundles?"

"That way oversimplifies the subject. The brain has 100 billion neurons with 100 trillion connections. It's an extremely complex organ – which is why it's so difficult to understand. But eventually, science will be able to describe the brain's electrochemistry well enough to be able to explain how emotions and motivations originate."

"I doubt that," says Deb. "Molecules don't fall in love."

"You don't know that for sure. The brain's limbic system has a set of structures like the hippocampus and amygdala, which seem to be involved in emotion."

"But the amygdala isn't going to get mad. Like me!"

"Couldn't the brain," says John, "and all the stuff inside, just be an instrument that is used by our spirit – just like a driver uses a car, or a pianist uses a piano?"

"That's pretty simplistic."

"Maybe, but I'm just a simple guy," replies John, "and it's hard for me to believe that a brain the size of a big grapefruit can be the source and repository of everything that I am."

"You know," says Mark, "I get all confused about this subject. So I do what most people do. I ignore it – block it out of my mind. That is, until someone I know dies, and then ... well ... I try to convince myself that we may indeed have an eternal spirit that lives on after our physical death. And when I do that, it's comforting – even though I don't completely understand how all that stuff works. You see Luke, the trouble with your evolved brain theory, is that it's so finite. If your entire essence is nothing but a multitude of interconnected neurons, then what happens to you when you die? All those nerve bundles just die along with you – and that's the end. All that is left of your evolutionary advanced brain is a bunch of dead, decaying cells. Your life is over. No heaven and no eternity. No future and no memory of the past."

"I'm afraid that's true."

"So, you don't think you will ever see your daughter again?" asks Deb.

I start to answer Deb's question, but I suddenly get a lump in my throat and I am only able to utter a quiet, staccato "No." But then I clear my throat and say decisively, "No, she's dead."

No one says anything else for a moment, until the awkward silence is broken by Mary, who says, "Matt, what do you think?"

"Well," responds Matt, "there is the binding problem."

"The binding problem?"

"Yeah, it's kinda hard to explain. Let's see ... ah ... Mark."

"Yeah."

"What happens when you see a beautiful woman?"

"That depends," says Mark. "Does she have any clothes on?"

"Oh, well ... let's say she's naked. What happens when you look at her?"

"OK ... ah ... lots of things happen ... ah ..."

"Let me help you out," says Matt. "Light reflecting off this beautiful, scantily clad body travels through the air and is projected upside down on the retinas in the back of your eye. Then what happens?"

"Well," says Mark picking up on Matt's train of thought, "there are those rod and cone things in the back of the eye that connect to the optic nerves that goes to the brain, and the brain turns everything right side up and puts together the images from each eye, and I see a beautiful picture of a naked lady in 3D. How's that?"

"Not very good."

"How come?"

"Well, the first thing you have to realize," says Matt as he stands up in front of the fire, "is that it's dark inside the brain. There's no light in there and there are no pretty 3D pictures projected on a big screen. Everything you actually 'see' is just an amalgamation of nerve impulses that somehow miraculously come together to form a picture in our minds."

"I never really thought of it that way before," says John.

"And the second thing you have to realize is that the picture that we eventually do see in our head is the result of some very complex processes. Initially, parts of the two optic nerves cross and share some information, but then they go to two separate sides of the brain. Most of the optic nerve endings terminate in the thalamus but then the activity begins to spread out – to the visual cortex, to the LGN, to the midbrain, and on and on. And before long, there's activity all over the place, and the remnants of your beautiful woman is spread out in a variety of locations all around your head. Humpty Dumpty has had a great fall."

"So how do you get Humpty Dumpty back together again?"

"Well, that's the binding problem. It's something no one understands, and many scientists think that the current laws of physics and chemistry are not going to be able to deal with this problem. We may need something new."

"But Humpty does get back together. We do see things."

"Oh yeah, but it may be more complicated than you think. Not only is the image of our woman reassembled, but our brains will also generate feelings that it associates with our final image, and the brain also pulls in selective memories of other events and other times – triggered, perhaps by our lady's long flowing hair or her seductive smile."

"I may not be looking at her face."

"OK, but you get the idea. The attraction you feel is somehow a combination of not only the visual, but also emotions and memories."

"Love at first sight."

"Or maybe lust. It doesn't really matter. The point is, emotions come into play and they merge with prior memories to shape our perception. And then some of our other senses, smell and touch and the memories of past smells and touches are all selectively collected from reservoirs inside our head. And finally, we see in our mind, a beautiful woman – constructed in part, from the disparate parts of our brain where the nerve impulses from the original image traveled, but enhanced with stored information retrieved from the memories of a soul that has known the euphoria and pain of past encounters. A soul that must now evaluate alternative choices, as it weighs the benefits of initiating a new relationship against the compromises that may have to be made because of prior commitments to others.

"You see, the final picture that we form in our minds is one that is the result of our consciousness. It is not just a picture of a pretty girl. It is an image filled with the anticipation of things to come, and saturated with memories of the past – all brought together through some miraculous mechanism that we call the binding process."

"Wow," says John. "You never really think about how all that comes together."

"Yeah, but that's what consciousness is. It's an awareness of everything around you and an integration of your feelings and experiences to help in interpreting what you see, hear, smell and touch."

"And it's not that little hippo-camp in your head doing all that stuff," says Deb.

"That's hippocampus."

"Whatever. You just have to recognize the fact that it's your soul – your eternal spirit, which makes you conscious."

"Maybe", says Matt, "but the subject needs much more study. There's still a lot we need to learn. Most of what has been written on consciousness is not worth reading."

"So science doesn't understand consciousness at all?" asks John.

"It does to a small degree, but some of the conclusions physicists have come to may surprise you," responds Matt.

"Like what?"

"We probably don't have time to discuss it tonight, because first we need to talk a little about quantum mechanics, and Schrodinger's cat."

"I just hope that that will be easier to understand than consciousness," says Nan.

“Actually, it won't,” replies Matt. “In fact, I'm sure it will be harder. You see, no one really understands quantum mechanics. Even Albert said that it was 'spooky'.”

“Really.”

“Yeah.”

“It's really spooky how this puzzle is coming together,” says Deb. “I think these brown pieces make some kind of house. See, here's the front door.”

“Albert didn't actually use the word 'spooky'. Did he?” asks Nan. “He probably used some fancy scientific term.”

“No, he said 'spooky', and tomorrow I'll tell you why.”

“This stuff may be way over my head,” says Mary. “I think I'll go to bed before I get any more confused.”

“Me too,” says Nan. “I'm tired. I'm sure I'll be UN-conscious in about 2 minutes.”

“You and me both. Good night.”

“Good night. See you in the morning.”

“Night.”

“Matt?” says John.

“Yeah.”

“Do you think sometime you can explain a little more to me about black holes – and what happens at the center – you know, where time stops?”

“Sure, it's really a fascinating subject.”

“Thanks.”

“Good night.”

CHAPTER 4 – DAY 3

Once you can accept the universe as matter expanding into nothing that is something, wearing stripes with plaid comes easy.

Albert Einstein

Morning Paddle

"Knock, knock. Hello. Anyone home?"

"Who's that?" I ask Nan as she jumps out of bed and peers out the window.

"It's Peter. I think he brought the motorboat. We'll be right down Peter," Nan shouts out the window.

We had reserved a small motorboat for the next few days to explore the lake and to ferry everyone back and forth to Chocorua Island – where we are having Nan's daughter's wedding. We expected the boat to be delivered today, but just not this early.

I quickly pull on a pair of shorts and a polo. Fortunately, it's much warmer today than yesterday morning and getting out of bed is not nearly as painful. I look at my watch and I am shocked to realize that it is almost 10 o'clock. Somehow, we had all slept through breakfast. "I didn't even hear the bell," I think to myself as I run down the stairs to let Peter in.

"Sorry to wake you," says Peter. "I though you would all be up by now. I just needed to bring you the keys for your boat."

"Thanks. Come on in. I can't believe we slept this late – and missed breakfast" I say as the rest of the sleepy heads in the family slowly mosey out of their rooms.

"It will probably do us good to miss a meal," says Deb.

"Yeah," says Nan, "but we will probably just eat twice as much for lunch."

"True."

"You're lucky you're just here for a week," says Peter. "The staff gorges themselves all summer long."

"So, how do you stay so fit and trim?"

"He's young," I say.

"Yeah, but I also try to get lots of exercise. I towed my kayak over here behind your motorboat. I have an hour off this morning and I thought I would paddle around Bennet Cove. You're welcome to come with me if you'd like."

"That would be great," says Mark.

"Can we take the motorboat?" asks John.

"John, that would defeat the purpose. Remember? We need EX-ER-CISE!"

"Yeah, I guess you're right."

Everyone quickly throws on some clothes and we meet on the dock, where we board our fleet of maroon and forest green watercraft, and line up behind Peter – who takes the lead. Peter knows he is guiding a crew of amateur boaters, and so he paddles at a slow, leisurely pace, so as not to outrun his older, slightly out-of-shape followers.

It's a beautiful morning on the lake – slightly overcast with the sun just barely peeking through the clouds and the haze – bringing a welcome warmth that begins to dissipate the cool morning mist floating just above the water. It's a gorgeous scene – with the trees, the rocks and the sky reflecting in the calm clear water. It reminds me of a late nineteenth century impressionist painting – lacking only a long white dress, a parasol and a picnic lunch.

We paddle up the cove and along a shoreline that is lined with large, majestic, granite boulders. Rocks that have survived centuries of glacial grinding resulting in irregular rounded shapes and odd stacked configurations – their worn faces and flabby abdomens betraying their age.

Tall, strong trees stand guard at the water's edge, with wide branches forming a protective canopy over our heads as we paddle close to the shore. We look up when we hear the distinctive laugh of a red bellied woodpecker and watch as he runs circles around the trunk of an old maple tree, stopping periodically to peck at a small unsuspecting insect. He pauses for a moment and gazes back at us, as if he is wondering who these strangers are in his backyard. But a moment later he flies away to another tree when he hears the rustle of another uninvited guest running through his branches – a small squirrel.

As we paddle along, we pass over giant submerged rocks just inches beneath the waterline. I think for a moment that our canoe will strike one of the shallow boulders, but we glide across the top with ease. I guess the clear water creates an illusion and the stones are actually deeper than you imagine.

The irregular lake bottom formed by the rocks provides a perfect hiding place for small mouth bass that are prolific throughout the lake. Many anglers come here in the summer just to fish. We may have to try our luck later this week because repeatedly, we hear a swooshing sound on the water, as a fish surfaces to gobble an insect that has inadvertently taken a morning swim.

"This is just lovely," says Nan.

"Shhhhh," whispers Peter as he stops paddling. "Look."

In front of us about 75 yards off our bow there is a large bird in the water. "What's that? A big duck?" asks Nan.

“No, no, says Peter. It's a loon. It's a water bird that is endangered in this area. We used to see them all over the lake, but development has encroached on their habitat and their numbers have declined. It's really a shame.”

“What a beautiful bird. Look at his plumage,” says Deb as we all admire his iridescent black head and his black and white checkerboard back.

“Loons,” says Peter, “are not like other water birds around here, like ducks, geese and cormorants. They are an entirely different species. Their closest living relatives are penguins and albatrosses. There are still quite a few loons that nest here on Squam Lake, but in 2005 the number of mating pairs dropped to an all-time low, and there is now a unified effort to bring their population back.”

Just then a motorboat comes up from behind and roars past our canoes and then veers off to the right, and passes about 30 yards from the loon. Peter waves his hand in an up and down motion, trying to get the boat to slow down, but the pilot appears to be oblivious to his signals.

“He's going way too fast this close to shore,” says Peter, as the boat's wake rocks our canoe back and forth. “Loons make their nests close to the shore because they are not very maneuverable on land, and motorboat waves often swamp the nests. It's one of the reasons for the declining population.”

The loon, watching the boat pass, flaps his wings and pulls most of his body out of the water as he makes a loud crazy laughing sound.

“That call he is making is called a tremolo,” says Peter. “It's used by loons to signal an alarm. It could be he has a nest close to here. Loons also make a call called a 'wail' that sounds like a wolf call. You'll usually hear it at night. Typically, the wail is used in social situations.”

“I remember that call from the last time we were here,” says Mark. “At first, we thought it was a coyote.”

We all watch intently as the loon swims slowly toward our canoes, but before he gets very close he dives under the water, and is gone.

"Where is he?" asks John.

"No telling, says Peter. "Loons are great underwater swimmers. They can expel air from their bodies and compress their feathers and glide effortlessly through the water."

"How long can they stay under?"

"Oh, about a minute. Keep looking around. He'll pop up again."

We all wait, continually scanning the water for what seems like two or maybe three minutes, and are just about to give up when John says, "There he is. Behind us. Five o'clock. About 40 yards out."

"How did he get all the way over there?"

"I told you. They're great swimmers. You can see ..." But before Peter is able to finish, the loon dives again.

"I wonder where he's going now," I say.

"Just keep still, and keep your paddles out of the water."

We all pull up our paddles and resume scanning the surface – until we hear Deb, who with a combination of surprise and wonder, says, "Ahhh."

The loon had surfaced about 25 feet to Deb's left. Close enough for everyone to see this enchanting bird, with its beautiful plumage and orange eyes that are staring right at us. We all sit motionless, returning the loon's gaze, mesmerized by his majestic presence and stately demeanor. We wait a few moments, no one moving, wondering what the bird's next move will be, almost in suspended animation. And after the loon has examined our fleet of watercraft, and he is confident that we do not pose an imminent danger, he slowly turns and begins swimming toward shore, and he makes a single, quiet 'hoot' sound.

"I think he's calling his mate," whispers Peter. And we all watch as another loon swims out from a small cove on the shore. "It looks like he's found his girlfriend."

"Loons mate for life. Don't they?" asks Matt quietly.

"We used to think so," responds Peter. "But now we have found that the mating process doesn't always go well, and loons will often find alternate partners."

"I can understand that," says Mark.

"Wow," whispers Peter. "Look. This is something you don't see very often. Look at her back."

We all look closely at the female loon swimming toward us from the shoreline, and as she comes closer, we see a small yellow beak in the middle of her back. It's the head of a baby chick. Probably just a few days old.

"Ahhhhh," says Nan. "How cute."

"The chicks can swim as soon as they are born," says Peter quietly, "but they usually ride on their mother's back for warmth and for protection from predators."

The mother and her chick swim over to her mate and then they both begin swimming up the shoreline. And we watch, motionless, as they move away.

"Well," says Peter. "I think I need to head back. I have to get to work."

We all turn our boats around and slowly paddle back toward our dock. About half way there Peter says, "You're on your own from here guys. I'm headed to the boathouse."

"Thanks Peter. That was a great morning."

"You're welcome. Let's do it again later this week."

"We'd love to. Thanks."

"How long until lunch?" asks Mary. "I'm starving!"

"Me too. I wish we hadn't slept through breakfast."

"Sorry y'all," says Mark. "But it looks like about a half hour before the bell rings."

"I don't know if I can make it that long," says Deb.

"Sure you can," I say. "We'll just sit here on the dock and pass the time talking about something interesting. Like loons or the weather ... or something spooky ... like quantum mechanics."

"Oh no. Not quantum mechanics. Anything but that," complains Deb.

"Well," says Mat, "we have a little time to kill, and I can give you a quick rundown of some of some fascinating experiments that have been done with very small particles in the quantum world."

"Here we go again."

"You know it's funny," says Matt, "but that loon this morning reminded me a little of quantum theory."

"How's that?" asks Nan, Matt's star pupil.

"Well," replies Matt, "you know how we saw the loon, and then he went under the water and we had no idea where he was until he surfaced again?"

"Yeah."

"Well, small particles like light and electrons are like that. You never are really sure where they are until they surface."

"Light is a particle?" asks Nan.

"Yeah. Albert said so."

"Of course."

"A particle of light is called a photon."

"Yeah, but how can a particle of light go through solid things?"

"Well, solid things are still mostly empty space, the gaps between most molecules are much larger than the molecules themselves – and photons are really small. In fact, how many photons do you think a 100 watt electric light bulb emits every second?"

"About a billion?" says John.

"No! About a billion trillion. That's a 10 with 20 zeros after it. They're really small."

"Wow."

"But studies have shown that even though photons are very tiny, they have a mass and they are affected by gravitational fields. And so, light is definitely a particle – just like an electron is a particle. Although, at times, they both behave strangely – unlike any particle you have ever experienced."

"OK, but what do you mean you are not really sure where they are until they surface?"

"Well, you see, in the quantum world, things don't really exist until you look at them. It takes a conscious observation to bring something into existence. Things only materialize when they are observed."

"Now that's spooky," says Mark.

"So, how does that work?" asks Nan.

"I'm not sure anyone knows how it works," says Matt, "but I can give you examples of experiments that have led scientists to some very strange conclusions."

"OK"

"There was a classic experiment that was performed many years ago, where a light beam was aimed at two small closely spaced slits that were made in an opaque barrier. The experimenters expected to see two bands of light projected on the wall behind the barrier, but what they actually found was something kinda strange. Instead of the two bands they expected – one band for each slit the light passed through ... they actually found a whole series of light and dark bands."

"I remember that experiment from physics class," I say. "You see multiple bands because sometimes light acts as a wave, and the particles passing through one slit interfere with the particles passing through the other slit, causing the light and dark patterns. Like the peaks and troths of two ocean waves coming together to form a whole series of larger and smaller waves."

"Yeah. That's true. But in later experiments, scientists developed the technology to fire one particle of light at a time toward the two slits."

"So that they couldn't interfere with each other?"

"Yes. And you know what they found?"

"Well, since the light particles can't interfere with each other, they probably found the two bands of light they were expecting to see originally."

"That's what everyone expected. But what they actually found was something very strange. Even though they only fired one particle of light at a time, they still saw an interference pattern – the same pattern they had seen before."

"Really?"

"Yeah."

"So what does that mean?" asks Nan.

"Well, you know, in the quantum world, the conclusions scientists come to are difficult to understand because they don't correspond with the way we think about the world."

"So?"

"So, the conclusion they came to was that each individual particle of light actually went through both slits at the same time, and interfered with itself."

"How could that be?"

"I don't know. That's just what happens. In fact, a small light particle can travel lots of different paths. It can travel across the room ... or do loop-de-loops and travel through both slits multiple times ... or it can take a trip around the block, ... or even fly off to a remote crater on the dark side of the moon before it goes through a slit ... and we don't ever really know where the light particle is or what path it is taking until we see it projected on the wall behind the slits. It's just like the loon when it's underwater – we have no idea where he is until we see him at the surface."

"To the moon?"

"Yeah. The probability that it actually travels to the moon and back is very small, but it's not zero. It can happen. And just like the loon underwater, we never really know where it is until it pops up. This little particle of light that we shoot through a small slit can take lots of different routes on its way to the back wall – and we have no way of knowing which path it is going to take."

"Well," says Mark, "why don't you just look beneath the water? Why don't you just watch the loon during his dive? Or watch each light particle of light as it passes through a slit? If you just look, you can see exactly what path the particle is on and where it is headed."

"Good question," responds Matt. "And that's exactly what scientists did."

"And?"

"And, they discovered that if you observe the light particle before, during ... or even after the particle passes through the slit ... for some strange reason ... it only passes through one of the slits – not both. And when you do this, when you consciously observe the particles, no interference pattern is produced, even though the only thing you did differently from the first experiment was to watch. And when you watch – when you consciously observe the light particles, all you see projected on the wall behind the slits are two solitary light bands – no interference pattern at all. You see, somehow, our conscious observation causes a light particle to materialize – to come into existence at a particular place in space and time. And when that happens, it can only pass through one of the slits in our barrier – not both."

"I'm not really sure I even care how many slits a photon passes through," says John. "Why does it even matter how things behave on microscopic scales? It really doesn't seem applicable. I mean, if I hit a tennis ball at two slits, the ball will always go through either one slit or the other."

"John," I say, "if you hit a tennis ball toward two slits, the ball is most likely going to go over the fence. Not through a slit."

"That's not funny!"

"OK, OK," says Matt. "But you see, you really do need to understand how things behave on microscopic scales to truly understand how things behave on the tennis court, because conscious observation is something we are constantly doing, and it affects everything in our macroscopic world – just like it does in the microscopic world. Let me give you another example. There's an experiment where you can fire a particle of light, or an electron, or an atom ... or anything else small ... at a semi-transparent mirror."

"What's that?"

"It just a mirror that doesn't work very well. Half the time the particle reflects off the mirror, and half the time it passes through. If it reflects, the particle is captured in a box. Say box A. If it passes through, it's captured in box B. Now, where is the particle?"

"Well, you only have one of two choices, so, I'll guess box A," says John.

"That's correct."

"Darn, I was going to guess box B," says Nan.

"That's correct too."

"What?"

"I know it seems odd," continues Matt, "but quantum mechanics tells us that the particle is in both box A and box B. And we can perform an interference experiment, just like the two slit experiment, that proves that is true. And those experimental results are unequivocal. The particle is in both boxes."

"I agree with Albert. This is spooky."

"I know, but you see, the particle is only in both boxes if we don't look. If we peer inside one of the boxes – if we consciously observe what is inside, everything changes. And when we look, the particle is no longer in both boxes, but it is in only one or the other. Somehow observation, conscious observation affects the experiment, and causes stuff to instantly materialize in a certain place."

"I do believe in spooks. I do believe in spooks," I say, imitating the Cowardly Lion in the *Wizard of Oz*.

"So," says John, "I don't understand. How can it be that our consciousness, something which is inside our brains ... affects things that are outside our heads?"

"Good question. I wish I knew the answer, but I have no idea. No one does. But experiments have led us to believe that our consciousness seems to be an integral part of the entire quantum world. And the quantum world is all around us."

"Really?"

"Yeah, and our consciousness, our self-awareness, our realization that we are alive and interacting with everything around us ... somehow, is also responsible for creating the reality that we experience."

"This is all really confusing."

"Yeah, I know. This confuses everyone. Even the scientists that have conducted these studies are confused. But no physicist will argue with the fact that observation, conscious observation, completely changes the results you obtain when experimenting with very small particles."

"It's still confusing."

"I know," says Matt. "But let me try to make it a little clearer by telling you about a thought experiment."

"What's a thought experiment?" asks Nan.

"It's an experiment you can't actually perform but where you try to extrapolate the outcome based on facts that you know."

"Alright."

"The experiment is called Schrodinger's Cat and it's similar to the two box experiment, except that one of the boxes is larger, and Schrodinger speculates what will happen when he places his cat in the box along with a Geiger counter and a corked bottle of poisonous cyanide."

"That sounds interesting," says Mark. "I guess the reason you can't perform the experiment is because of objections from animal rights groups."

"No, actually the reason is that it is too difficult to isolate the cat box from any conscious observation. It's just too big, and our current technology doesn't allow it."

"OK, so is the cat going to die from cyanide poisoning?"

"Well, that depends. You see, the experiment is set up so that if we shoot a small particle and it enters the box, the Geiger counter fires and pulls the cork out of the cyanide bottle, and the cat is toast."

"Poor kitty."

"But remember, we fire the particle at a semi-transparent mirror, and there is only a fifty percent chance that the particle enters the cat box."

"So we have to look to see if the cat makes it."

"Yeah, but that's the whole point. If the cat box is completely isolated from all conscious observation ... the particle we fired is in the cat box AND the other box at the same time – both box A and box B. And the Geiger counter in the cat box is both fired AND unfired. And the cork is both pulled AND un-pulled. And the cyanide is both released AND not released. And the helpless cat is both dead AND alive. And here's the really amazing thing. It's not until someone actually looks – until someone makes a conscious observation – that something really happens."

"So what does happen?" asks Nan.

"Well, let's say we wait a few hours before we take a look in the box. When you finally do open it – at the very moment that you look, the Geiger counter will change from being both fired AND not fired ... to fired OR not fired. And the cyanide bottle cork will change from being pulled AND not pulled ... to pulled OR not pulled. And our brave cat, instead of being both dead AND alive ... will either be dead OR alive.

"I know it's hard to understand, but the reality that we see when we open the box ... is a reality that is created by our conscious observation ... and it happens at the very moment that we look."

"Now that's really spooky," says Mark.

"So what does all that mean?" asks Nan. "How does that affect us?"

"We are not really sure. But it does suggest that consciousness is not just something that goes on in our heads in a bunch of nerve bundles. Consciousness is a fundamental property of the universe, and somehow ... is responsible for creating the reality that we experience in our daily lives.

"So, you don't believe that consciousness is just a property of an evolutionally advanced brain?" I ask.

"No, I don't. Because you see, consciousness is not confined to the nerve bundles in our cranium. Consciousness travels. Somehow, through some mechanism we don't understand, our consciousness communicates with all the very small particles in the world around us. And because everything is made of those small particles, the inference is that our consciousness also links us to all the big things as well."

"Like the cat in Schrodinger's box?"

"Yeah. And everything and everyone else. Every interaction we have with the world ... involves consciousness. It's not just a quality of neurons firing sequentially in our heads. It's much more than that."

"So," says Mary, "how does all this relate to the binding problem we talked about yesterday? You know, where the brain pulls together the things we see and feel and remember and somehow creates our consciousness – our awareness of the things going on."

"Well, you see, the binding problem now becomes much bigger. Not only do we have to explain how our consciousness functions in our heads – we also have to explain how our consciousness interacts with our surroundings. And as of today, scientists have no idea as to how any of that works."

"You know," says John, "I had an employee in my flower shop in the Bay Area, and whenever she would talk about or even think about her mother ... for some strange reason ... her phone would usually ring. And when she answered it, inevitably, it would usually be her mom on the other end of the line. It was uncanny."

"That's just coincidental," I say. "Her mom probably calls her ten times a day."

"Maybe. But maybe your consciousness can travel outside your brain. And maybe there can be a connection – a consciousness connection between people – even over long distances – even as far away as the moon."

"I doubt it."

"Could that be how prayer works?" asks Mary.

"What?"

"You know, prayer. Could it be that somehow my consciousness ... my desires ... my prayers ... can travel as well? You know ... outside my head ... great distances ... all the way up to God in heaven? And perhaps that's how God can understand me ... and how He can know my thoughts, and understand my innermost feelings."

"That's just silly," I reply.

"I don't think it's that silly," says John. "Many of my friends in California believe in a universal consciousness – where everyone is connected in some way to everyone else – and especially to the people they love."

"That's a real good reason not to live in California," I reply sarcastically.

"We'll, I'll admit that some of them go a little too far," says John. "But getting back to Mary's point, I have often wondered how God can be everywhere, and how He can respond to our wants and needs. And maybe this consciousness thing is somehow involved in that communication."

"That's pretty far-fetched."

"Yeah," says Matt, "it is. But you have to realize that in the world we live in, things are not always as they appear. And you have to be open to accepting the fact that we live in a strange place – a place we really do not fully understand."

"Can we go to lunch now?" asks Deb.

"I don't think the bell has rung yet," says Mark.

"Maybe we can go up and ring the bell ourselves."

"Yeah," says John, "but if we don't consciously look at the bell, the rope might be both pulled AND not pulled – and the bell may be both rung AND not rung."

"That's a good illustration of the problem we have performing this type of experiment with big things," says

Matt. It's impossible to isolate large objects from a conscious observation."

"Because of all the little kids in line waiting to pull the rope are looking at the bell?"

"Yeah, but also because the bell gives off heat in the form of infrared radiation – heat that we can feel. And the bell emits odors because of the dust and pine sap that has landed on it – which we can smell. And the bell's gravitational attraction, although tiny, can affect our sense of space and time. There are lots of ways for ..."

"Let's eat," interrupts Deb. "I hear the bell."

"Thank God someone looked," says Nan, as we all head off to the dining hall.

"Let's go over to the island," says Nan after lunch. "I can't wait to see what it looks like."

"I think I'm too full to go right now," I say.

"How many reuben sandwiches did you eat anyway?"

"I love those things so much."

"So how many?"

"I lost count."

"OK," says Nan. "Just suck in your gut. We're headed to the island."

The island Nan is referring to is Chocorua Island, also known as Church Island, where we will be having the wedding tomorrow. Nan has never seen the island and is anxious to preview the place I have talked about so much.

"Hope this motor starts," I say as I help Nan into the boat. Then I pump the bulb on the gas line, pull the choke and turn the key – and the motor immediately revs up with a slight vibration and a puff of smoke.

"I'll untie the lines," says Nan. "Is it far?"

"Oh no. It's less than three miles away. We'll be there in about five or ten minutes. That is, if I remember how to get there, and if we don't get lost, or hit any rocks."

"Just be careful."

"OK"

Chocorua Island has a long and interesting history dating back to the late 1800's. In the fall of 1881 Ernest Berkeley Balch purchased the island for forty dollars and established Camp Chocorua, the first resident summer camp for boys in the United States. Summer camp was a new concept at the time, and was slow to become an accepted activity because sleeping out in the open air was thought to breed and incubate a number of fatal diseases.

But it wasn't long before the idea of "Let's-get-the-kids-out-of-the-house-for-the-summer" caught on, and the camp was filled with boys between the ages of eight and fourteen – most coming from well-to-do families throughout the northeast for the adventure of their lives. And what fun those kids had – fishing in the pristine waters – building their own boats to sail from island to island – swimming and bathing in the cool lake. A Robinson Caruso dream come true. And best of all, no school books were allowed. Instead, the boys were consumed with activities. They spent most of their mornings doing chores around the camp that earned them an allowance of twenty-five cents per week, and then they spent the rest of the afternoon exploring. A perfect summer diversion for young adolescents with enough pent up energy to drive their proud and proper nineteenth century mothers bonkers.

On the northeast corner of the island, in a grove of silver birch trees along the shore, the boys themselves built an open air chapel for their Sunday worship services, which featured, as you might expect, an angelic boys' choir. The beautiful trees not only provided a perfect setting to contemplate the majesty of creation – but they also provided seating, supports, fences, trail markers – and a ten foot cross, which was erected in front of the altar, overlooking the lake.

The outdoor chapel is the only surviving structure from the camp, which closed in 1889 because of the death of the founder's sister, and various financial difficulties. The island was eventually sold to Clinton Hoadley Crane a twenty-one year old former camper. Crane later deeded the land to the Chocorua Chapel Association which was incorporated in 1903 for the purpose of conducting church services during the summer season on the island – which it has done every year since.

"This is lovely," says Nan as we pull up to the dock. "Watch out for that big rock."

"I see it."

We tie up to the dock and walk along the winding dirt and pine needle path to the far side of the island toward the chapel. "It's been forty years since I have been here," I think to myself, "and not much has changed."

When we enter the back of the chapel, it feels as if time has stopped and I marvel at how this small piece of land has resisted a culture that so admires technology and advancement. "It's so simple." The altar is a big rock with wild flowers growing around its base. The lectern is another appropriately shaped stone, and the beautiful lake in the background is more inspiring than any stained glass window I have ever seen.

"This is perfect!" says Nan.

There have been some minor improvements to the chapel over time. The benches have been upgraded, and some stones moved and some of the brush cleared, but not much else has changed. The simple cross in front is still made of two tree limbs with a couple of nails holding them together. I guess that's appropriate.

During its first forty years of operation, most of the services at the chapel were conducted by the Reverend Clifford Grey Twombly, an Episcopalian minister and diehard fisherman who spent his summer vacations on the lake. The Reverend was an avid reformer in his home town of Lancaster Pennsylvania, and he would reel in the owners and patrons of barrooms, gambling houses and brothels with the same fervor that he reeled in small mouth bass. And very few got away.

The Reverend Twombly preached his final service in 1942, just before his death. Since then, the chapel association has arranged for guest ministers from all around the northeast to preside over summer Sunday services.

“You know, Luke. We should have come to the service here yesterday. It might have brought back some good memories.”

“You would go to a Christian church service?”

“Sure. Why not? Who knows? I might be attending a service in a mosque someday.”

Nan picks a couple wild flowers from the side of the trail and holds them in front of her with both hands. Then she begins walking down the center aisle, slowly, one small step at a time.

“I can't wait to have the wedding here,” says Nan. “You were right. It is beautiful.”

The first wedding in the chapel was held in 1916, when Miss Margaret Armstrong married Mr. Arthur Howe, an All-American quarterback from Yale.

Since that time the chapel has become a popular place to tie the knot. So much so that in 1974 the Chocorua Chapel Association passed a proposal requiring an application be submitted and approved before a wedding would be allowed. Nan and I completed the application several months ago, and received approval within a few days. I think one of the reasons we were so readily approved was because I noted on the application that my family and I attended services here as kids, and that one Sunday, my father led the singing and collected the offering while we were waiting for the pastor – who was delayed because his boat sank after filling with water during a Saturday night rainstorm.

“Where's the organ?” asks Nan.

“Over here. I'll show you.”

I lead Nan over to a small wooden building just to the left of the altar. It has two large doors in front that are latched closed with a slightly rusted padlock. “It's in here. All you have to do is swing the doors open, pull out the organ bench and start playing.”

"Is the pump in there too?"

"Look over here," I say as I walk around the small fence on the side of the building. "This thing doesn't even require any electricity. The bellows are run by a small crank that fits into this hole – here in the side."

"Is it hard to turn?"

"I'm not sure. Seems like every time I was here, there was always a long line of young boys hoping to be selected to be the one to crank the organ – and I never got a chance."

"I think that Ruth has some friends with young boys that are coming to the wedding. We can let them have the pleasure."

The organ was donated to the chapel in 1935, in the middle of the Great Depression, and has served for decades to lift the spirits and warm the hearts of everyone attending the summer Sunday services. The beautiful instrument has had to be refurbished a few times over the years, but it has survived admirably, its main advisory being field mice who like to nest in the pipes.

"Hopefully, this thing will work tomorrow."

"Yeah, and I hope someone remembers to bring a key to open the lock," says Nan.

On the way back to the camp we stop at Birch Island where we once picked wild blueberries as kids. I remember coming here years ago, trying to scavenge for at least two heaping cups of blueberries – enough for the Deephaven cook to be able to bake us a pie for desert. Just for our family. Probably the best pie I have ever had.

There are no docks on this island so I motor over to the leeward side and turn off the motor about 15 feet from the shore and paddle the rest of the way in – avoiding rocks on both sides.

Nan jumps out of the boat and walks along the shoreline. "Doesn't look good for blueberries," she says. "Lots of bushes but all the berries are green. Must be too early in the season."

"You may be right. We usually came here later in the summer."

"Look!"

"What?"

"Behind you. I think it's an eagle."

I look behind me and I see a beautiful, white headed bald eagle, flying a few feet above the water. He then climbs higher, circles once around the island and heads off to our right.

"I don't think I have ever seen a bald eagle in the wild," says Nan. "Where do you think he's going? Let's see if we can follow him."

Nan jumps back into the boat and pushes off as I crank the motor. In our haste we come precariously close to a large boulder on the port side but I am able to avoid it by a few inches.

"I think he landed on that island over there," says Nan as we motor full speed toward the east.

As we approach the island we realize that we are not the only ones on the lake that know about the eagle. There are about half a dozen boats anchored just off the island, with kids and binoculars and cameras – intently watching a large twig nest in one of the treetops.

"Looks like we are probably the last two people on the lake to find out about the eagle nest," I say as we watch the magnificent bird sitting on one of the tree branches – with his beautiful white head glistening in the early afternoon sun.

"I love this place," says Nan. "I feel at ease here. It seems like home – even though I have never been here before."

"Yeah. I know what you mean."

"We need to make plans to come back next year. OK?"

"Sure. But right now we need to get back to the camp. You have to leave."

"Yeah."

Nan is planning to drive into Holderness this afternoon to meet up with her daughter and many of the wedding guests who are arriving today. Nan said that I was welcome to join everyone and spend the night with her at the Manor Hotel where most of the guests are staying, but I declined. Nan's x-husband and I never have gotten along very well, and I didn't want to put a damper on any of the pre-wedding activities. Besides, I really don't know anyone, and I have only met Nan's daughter on a few occasions.

"Here we go," I say as I bring the boat up on a plane. "We'll come back here later this week, after the wedding."

"Can't wait."

"You sure you can drive this monster?" I ask as I give Nan a boost into our four-wheel-drive pickup truck cab.

"Sure. No problem."

"OK. I'll put your suitcase here in the front seat. It looks like it may rain this afternoon, and I don't want it to get wet in the back."

"Thanks."

"You sure you packed enough clothes?"

"I'm only staying two nights, Luke."

"It was a joke, Nan."

"Oh. Bye. Love you."

"You too. Bye. See you tomorrow at the wedding."

Nan backs the truck out of the parking space, narrowly avoiding a granite boulder on right side, and is about to pull away, when she stops for a moment and rolls down the passenger side window. I look in and say, "Forget a pair of shoes?"

But Nan just gives me one of her serious looks – a look I have seen before. It is a look that I don't like, because I know the tenor of the conversation that usually follows that look. It almost always leads to a discussion about my failings and inadequacies, and about my ill-fated attempts to improve my social interactions. And as I stand there, hunched over on the window ledge, I realize that I was right about the shoes. And I wait for the first one to drop.

"Luke ..."

"Yeah," I reply.

"Please try to be sociable at the wedding. It's a big day for my daughter, and it's important to me that everyone have a good time. And I don't want you to mess that up."

"OK," I reply, knowing that short simple answers are the best.

"And maybe, just maybe, you could try to be nice to my ex-husband."

"OK. I'll try."

"I know you don't like him very much, and I know the last time you were together you said that you never wanted to see him again. But you need to make an effort to be civil."

"OK."

"You know, it's his daughter too – and I just want to be sure that this is a happy occasion for the whole family. Everyone. Despite all the obstacles."

"OK," I repeat, as Nan rolls up the window and drives away.

Lately, these conversations have been occurring more and more often. I guess Nan's patience with me is growing thin. But that's probably justified, because my progress has been slow. And I know it.

I want so badly to come out of the hole I have dug for myself. To be the happy, gregarious, socially active person I know I can be. The person I once was. But it is so hard, and I can feel Nan's frustration mounting as I continually revert to behaviors that she deems unacceptable.

It's been seven years since we were married. The first five years after the wedding were the happiest times I have had since my daughter's accident. But recently, storm clouds have been brewing, and I can see, clearly, that the turbulence is increasing, and that sunny carefree days of the past are being obscured by an increasingly ominous sky.

And I fear the impending storm.

I walk back to the cabin and go upstairs to change into my swimsuit. It is an unusually warm afternoon and John was talking about jumping into the water. I'm not sure I'm that brave, but I think that I may dangle my feet off the dock and take a few pictures.

I begin changing in front of the bedroom window, knowing from experience that it is difficult to see anything inside the cabin from outside, because of the wire screens on the windows and the bright afternoon sunshine. We used to pull the flimsy, thin curtains closed when changing, when modesty was an issue, but time, wrinkles and a few fat rolls have ameliorated the embarrassment of being seen without any clothes on.

"Maybe," I think, "it should be the other way around. Maybe I should be more worried about being seen naked now – with my receding hairline and an expanding gut – as opposed to years ago, when I had a full head of locks and six-pack abs. Oh well. I guess that's just part of getting older."

It seems the most painful part of aging is looking in a mirror – when your idealized self-image is confronted face to face with the reality of a deteriorating physique. It's so easy to envision yourself as an invincible twenty-five year old, and try to ignore the body's obvious, steady, downhill decline. I guess that's why they don't put mirrors on nude beaches – where all the older and overweight middle-aged-has-beens parade around thinking they still have something to show off.

"I wonder what I would look like in one of those little stretchy Speedo bathing suits." I think to myself.

I look out of our bedroom window and see Mary, Deb, Matt, Mark and John all sitting on the dock. It's a great view from up here on the second floor of the cabin. I can see a wide expanse of the lake, with the western mountains in the background, where long billowing columns of cumulous clouds are beginning to grow and merge together.

The weather on Squam Lake can be very unpredictable, but even a novice like me can see that we are probably destined for an afternoon shower. The camp always posts a daily weather forecast at the main office, but all the regulars know not to pay any attention. The standing joke is that the Deephaven weather girl uses a dartboard to determine the "chance of precipitation" and the afternoon "high".

The weather on the lake can change in a matter of minutes. It doesn't take long for a calm bright sunny afternoon to be transformed into a windy torrent of blistering rain as weather systems wind their way through the local hills and valleys. Mt. Morgan, the tall peak just to the west of Squam Lake sometimes obscures approaching weather, and boaters are often stranded on the lake when storms approach unexpectedly.

As I am pulling up my bathing suit, I notice that there is a message on my cellphone. "That's odd," I think to myself. "I didn't think we had coverage here."

The cellphone service around the camp is very spotty. Some places, like the main dock, are better than others, but usually your phone just displays the "No Service" icon. However, for some reason, right now, I have two bars.

My hand is shaking a little as I pick up my phone. The message is from my doctor's office and I know why they called – to give me the results of my cancer scans. You would think that I would have gotten over the nervousness my now. After all, every time they call, it's the same result. "Everything's fine," they say. "You're cancer free." And then they tell me to come back in a year or so for another scan – as if "cancer-free" really means "cancer-we-can't-see-that-is-hiding-somewhere-just-waiting-to-attack."

I dial my doctor's office, not bothering to listen to the message they left on my phone, because I know they never leave test results in a voice mail. I can feel my heart pounding as the phone rings. Three, four, five, six rings ... and then, "Hello, Dr. Plank's office."

"Yes, this is Luke Dawkins. You have some test results for me?"

"Just a minute please."

And then after only a few seconds, I hear, "Luke, this is Doctor Plank. I'm glad you called back."

And I know immediately that this is not going to be good news. I usually have to wait a lot longer, and I never speak to the doctor. It's always a nurse. "Hi," I say apprehensively.

My doctor, who is normally very cordial, dispenses with the usual social formalities of "How are you doing?" and "How's your vacation?" and gets right down to business. "It looks like there may be a problem. We are going to have to do some additional testing. You have three new tumors that seem to have ..."

Then I hear two small beeps, and the call drops ... and so does my heart ... and my hopes ... and my dreams. I hold up my cellphone and try to steady my hands to redial the number, but the display just says, "No Service".

I knew there was something wrong with me. I knew there was a reason I was having difficulty carrying suitcases and playing tennis, and I somehow knew, in the back of my mind, that someday this cancer would return.

But why does this have to happen to me? And why now? Why is it that every time my life seems to be headed in the right direction, God beats me down and delights in my pain?

I frantically walk around the room, pacing from one corner to the other, desperately trying to pick up a cell phone signal. Until finally, over next to the window, I get one bar. I quickly redial the number for my doctor's office, but after the phone rings twice, I hang up. "It really doesn't matter," I think to myself. "I don't need to talk to the doctor. At least not today. I know what he is going to say. He is going to tell me that my cancer has returned but that I should not be discouraged. He'll say that there are new treatments available, and that we can 'lick' this thing just like we did before."

But I know better. I know that when lymphomas return, they return with a vengeance, and survival rates are not good. And as I sit in the rocking chair in front of the window, watching everyone on the dock enjoying an afternoon on the lake, I realize that this may be the last time I ever come to this beautiful place with Nan, and with the rest of my family. And I have to wipe my eyes with my sleeve to keep tears from running down my cheek.

It seems like hours have gone by as I sit in front of the window, rocking back and forth like a psychiatric patient – worrying about how advanced my cancer is ... wondering how much time I have left ... and watching outside as the storm clouds continue to grow.

I begin thinking in clichés, "Why can't I live happily-ever-after, like everyone else? Why have I been dealt such a bad hand? Maybe it's just time to throw in the towel, and admit defeat. No, I just have to stop feeling sorry for myself – pull myself up by my bootstraps. Maybe there's a silver lining in this cloud – somewhere. I just need to take it one day at a time. Perhaps we can 'lick' this thing."

And with a little self-encouragement and a little less self-pity, I begin to feel a little stronger. I lift my head higher and peer out the window at the lake below and I am strengthened by the warm refreshing afternoon breeze.

From my higher vantage point, I can see and hear almost everything going on outside. John has been trying for some time to muster the courage to jump into the frigid water and the rest of the family is encouraging him.

"You're a big chicken, John," says Mary.

"Yeah, cluck, cluck," adds Mark.

"I don't see any of you jumping in with me," says John.

"Yeah," says Matt. "But you made a commitment. You said you were going swimming, and so, you have to follow through – before it starts raining. Just jump!"

"I can't."

I look up at the sky and see the dark clouds coming closer. Constantly growing – strangling the sunlight – choking the life out of the day. Just like a cancer, they gradually and steadily take control. Relentless in their advance. Unstoppable.

And then, I hear a big splash. For a moment I think that I have missed John's big plunge, but when I look down, he's still on the dock.

Then I hear a loud "Woooooo Eeeeeee." It's Stephen, our next door neighbor. "Come on in John. You can do it," he says as he begins swimming out to the floating platform in front of his cottage.

"OK, OK. I'm going to jump. Just give me a minute."

John procrastinates a while longer. Then he starts to jump. Then steps back. Then starts again and backs off once more.

"Just do it!" shouts Mark.

Stephen climbs out of the water onto the floating platform and yells, "OK let's go! I'm waiting!"

The wind has started blowing harder and John realizes that the weather is deteriorating fast and that he is running out of time. So he takes a few steps back, gets a running start and jumps – feet first – screaming the whole way. "Ahhhhhhhhh!" A few seconds later his head pops above the water, and he continues, "Ahhhhhhhhh. It's cold!" He then gathers himself and attempts to ward off hypothermia by swimming out to our floating platform, directly in front of our dock, about a sixty feet from the platform where Stephen climbed out of the water.

"Way to go!" shouts Stephen from across the water. "I knew you could do it John."

"That was painful," responds John.

About that time a huge wind gust races across the lake and the trees rustle a warning as their branches begin swaying back and forth – doing their best to sound an alarm as the sky is becoming increasingly black and the tops of the waves on the lake are turning white. I look up toward Mt. Morgan and see the impending storm, sneaking out behind the summit and headed straight toward our cabin.

As a few raindrops begin to fall, tossed around by the swirling wind, I hear the screen door from the back porch open and Deb runs out. I guess she must have gone inside earlier to work on the puzzle.

"Get inside! Now!" shouts Deb in a panic as a steady, light rain begins to fall. "Hurry! Hurry! There's not much time! Get inside!"

And just then ... at that very moment ... a blinding bolt of lightning crosses the sky, and the immediate ensuing thunder rattles the cottage, with a loud deafening clap that reverberates through the room.

I am startled for a moment. "That was so close," and I look around the room thinking the lightning might have struck the house. And then I look outside, fearful for everyone on the dock. And I start counting people. One, two, three four, five. Matt, Mark, John, Mary and Deb all seem to be OK, except that John is still standing on our floating platform in a very vulnerable position. For a moment I don't understand why no one is moving – running for shelter. Everyone is just standing still, motionless, staring out over the lake, as another bolt of lightning crosses the sky, with another blast of thunder.

But then I look over to the right and I see something that causes my entire body to quiver. It's Stephen ... floating face down in the water ... sinking slowly as the intensity of the rain increases.

I don't know what happened – whether the lightning hit Stephen or perhaps hit close to Stephen. Or maybe, somehow Stephen fell and hit his head when the lightning flashed by. But it doesn't matter. His head is under water and he's not breathing. He needs help or he's going to die.

But I can't stop shaking, and once again I get that feeling. That overwhelming fear of death feeling. The same feeling I had in the tunnel ... the same feeling I had when my parents died ... the same feeling I had after the car hit my daughter's tricycle.

I try to jump up out of my chair, but I can't. My limbs are completely unresponsive and all I can do is stare out the window as my body trembles uncontrollably. And once again I am paralyzed, and once again, when faced with an emergency – a life and death situation – a situation that requires action, I am unable to respond. Unable to help. Worthless.

Everyone else is frozen as well, apparently trying to evaluate the risk of initiating a rescue operation. But, after a brief pause, almost simultaneously, the entire family jumps into action – believing that saving Stephen's life far exceeds the perils of mounting a lifesaving effort in the middle of a thunderstorm – and possibly thinking, subconsciously at least, that lightning never strikes twice in the same place.

What happens next you can only liken to a Keystone Cops free-for-all. It would have made a great comedy routine, if only the situation had not been so tragic. So, as the rain begins to fall harder than ever, and another bolt of lightning strikes across the cove, Mark, Matt, Deb and Mary begin a heroic, though ill fated, rescue attempt.

Mark jumps in the motorboat but is unable to get the motor started because he forgets to pull out the choke. He keeps cranking the engine as Matt grabs a paddle and attempts to make a fast entry into a kayak. It's a gallant and athletic effort, but he makes a slight miscalculation and loses his balance and falls into the water, swamping the kayak beside him. Mary and Deb climb into a canoe but just as they are about to clear the dock another huge gust of wind hits and causes the sailboat to capsize, with the mast falling on top of the canoe. Deb tries to push the mast off, but pushes a little too hard and the canoe turns over as well, dumping Mary and Deb into the frigid water.

By now the heavens have opened and it is raining so hard that it's difficult to see Stephen. He's just a blur on the water, with gale force winds and thunder and lightning all around.

Mark finally gets the motorboat started and decides not to wait for Matt, Deb or Mary who are just climbing out of the water, and he takes off from the dock at full throttle. But, in his haste, he forgets to untie the stern line which is secured to one of the dock's supporting posts ... and the boat's momentum pulls the post completely off its footing on the lake bottom. That causes the end of the dock to collapse, along with Matt, Deb and Mary, who once more fall into the water. The weight of the supporting post and the dock pulls the rear of the motorboat underwater for a moment, and the motor stalls as the boat fills halfway with water.

There seems to be little hope for Stephen, and I watch as his lifeless body slowly sinks below the surface of the water. "It's over," I say out loud over the roar of the relentless wind and rain. "He's dead."

But John, still standing on the floating platform shivering from cold and fear, decides to make one last ditch rescue attempt. He dives into the water and frantically swims toward Stephen attempting to make a solo water rescue of a man that probably outweighs him by about a hundred pounds. John is not a strong swimmer and is tired and out of breath by the time he is able to reach the area where Stephen went under the water, but he summons his last bit of energy, and begins searching underwater for some glimpse of Stephen's sinking corpse. Although the visibility is very limited above, John is able to see Stephen in the clear water about two feet below the surface, and manages to bring him back to the top by grabbing his hair and pulling with his last once of strength. John is too exhausted to do much else, but he is joined in a short time by Matt on the paddle bike and together, somehow, they manage to get Stephen on top of one of the bike's pontoons and over to the shore, where Deb and Mark begin CPR.

I am still sitting in my chair, trying to control my breathing and my trembling, and I begin to wonder how long Stephen had been in the water, but it's hard to estimate. Everything happened so fast. I try to recall how long someone can go without oxygen before there is significant brain damage, but I can't remember.

"... one and two and one and two and one and two and ... breathe Stephen, breathe!" pleads Deb as she compresses Stephen's chest. "One and two and one and two and ..."

"Here, let me take a turn," says Matt. "One and two and one and two ..."

By now the rain has let up a little but there's still a steady downpour. Stephen's wife comes running down to the shore, screaming in disbelief, and the five teenagers follow. All six of them are all distraught – crying, shaking, sobbing, pacing back and forth, not knowing what to do. Until, under the direction of one of the younger girls, they all kneel into a circle and begin to pray.

I understand their grief all too well. I know what it is like to be a helpless bystander – watching someone you love slowly pass away – unable to do anything to change or impact the impending events – praying to God as a last resort for help – help that you know will never come. There's a pain in your soul that emanates from deep inside. It's a pain that is difficult to describe, but a pain you never forget ... because you feel it every day for the rest of your life. And as I witness these tragic events, the pain that I have borne for so long becomes more acute, and I begin to cry.

A few minutes later an ambulance pulls up from the Holderness Fire and Rescue Department. I don't know how they got here so fast, or who called them, but they are a welcome site. They quickly take control. Stephen is apparently still alive, thanks to the valiant efforts of our family – at least everyone except me.

And as they carry Stephen off on a stretcher I begin to get some strength back in my legs and I walk over to my bed where I collapse. Exhausted, completely drained emotionally and physically – even though I have done nothing.

The dinner bell begins to ring as Matt, Mark, John, Deb and Mary come back into the cabin. I lie still in my bed, not wanting to let anyone know where I am. Preferring to be alone for a while – trying to figure out some explanation for my irrational behavior – my paralysis – my uncontrolled trembles – my fear of death. But it is of no avail. I have no answers.

I can hear everything being said downstairs, through the cracks in the walls, and the knotholes and the open windows. And I can feel everyone's' exhaustion as they speak in short, slow sentence fragments.

"Not really hungry," says Mary

“Me neither,” says Mark.
“Ought to eat something,” says John.
“Guess so.”
“Ah-Choo.”
“Bless you.”
“Thanks”
“Better change.”
“No showers?”
“No time.”
“OK”
“Be right there.”
“A few minutes?”
“Yeah.”

By now the rain has almost stopped and the cabin is quiet for a few minutes as everyone changes into dry clothes without saying much. Until on the way out the front door I hear Mary ask, “Anyone seen Luke?”

“No. Not all afternoon,” replies Mark.

“Maybe he changed his mind and decided to go with Nan into town.”

“Maybe,” says John, “but you think he would have left a note.”

“Yeah, well maybe we'll see him ... ah choo ...”

“Bless you.”

“Thanks. See him at dinner.”

“Yeah.”

I am still lying in my bed, almost asleep when everyone returns from dinner.

"Mark," says Matt, "if you could mix me up one of those double scotches of yours, I'll start the fire."

"Sure, I think we could all use a double something tonight."

"Make mine a triple," says Deb.

"No problem."

Mark begins mixing drinks while everyone settles in around the fireplace, and after a few minutes, I can hear the crackling of the fire and the squeak of the floorboards as the chairs begin to rock back and forth. No one says much for a while – just staring into the flames I presume, until John breaks the silence.

"Do you think he's going to make it?"

"I don't know," replies Deb after a few moments. "Honestly, I thought he was dead before you pulled him out of the water. Did anyone see whether or not the lightning actually hit him?"

"No."

"No," says Matt, "but his toes were burned."

"They were?"

"Yeah. So, even though it may not have been a direct hit. He had a good amount of electricity running through his body."

"Sad."

"Yeah."

And after another period of silence, John asks, "Deb, you knew something bad was going to happen. Didn't you? You know, when you ran out of the house ... just before the lightning strike ... screaming to everyone to come inside."

"Yeah."

"But how did you know? Was it just the storm clouds and the wind and the rain?"

"No," says Deb.

"One of your feelings?" asks Mary.

"Yeah. I knew there was danger."

"You know," says Matt. "I have struggled to understand Deb's premonitions for years now, and I'm still confused. At one time I thought that many of Deb's feelings could be attributed to coincidence, but I don't believe that any more. Coincidences don't happen with regularity, and even though Deb's feelings are not always correct, I have learned that it is prudent to pay attention."

"I like it when you pay attention to me darling," says Deb.

"I know sweetheart."

"You think it might just be ESP? You know, some type of extra sensory perception?" asks John.

"I did for a while," says Matt, "but I soon realized that there was much more to it than that. And the only reasonable explanation I have been able to come up with ... after many years of marriage ... is that something is going on in Deb's head that our current understanding of science ... just can't explain. You see, it seems as if Deb has some type of strange connection – a connection that allows her mind ... her consciousness ... somehow ... to sense events around her that no one else can see. And sometimes ... she can even sense things that are about to happen in the future."

"Yeah, sometimes I can," says Deb. "But it really doesn't happen very often, and I'm not always right."

"But it does happen periodically," says Matt, "and I have no idea how it works. And so the only explanation I have been able to come up with, is that Deb has this connection to another place – a place possibly hidden in some obscure dimension of space and time. I know this sounds a little flaky, and I know this is not a very scientific explanation. But right now, it's the best I can do."

"Matt," says Mark, "it must make life really difficult. I mean, it must be almost impossible to keep a secret from Deb. I don't know if I could live like that. You know, like, what happens if you meet a beautiful young lady and you want to have an affair?"

"He doesn't even have a chance to get started," says Deb. "I keep really close tabs on him. Mark, you need to find a girl just like me."

"That would never work," replies Mark.

"So Matt," says Mary, "Do you understand anything at all about this connection that Deb has, or do you have any thoughts about what she's actually connected to?"

"No, I wish I did, but I don't ... but there is something there ... and it's real."

"Do you think that maybe Deb just has a higher level of that wondering consciousness thing we were talking about earlier?" asks Mark.

"I don't know for sure. But ..."

"You know, it should have been me," says John abruptly changing the subject.

"What do you mean?" asks Mary.

"Don't you see? It should have been me. If we hadn't changed cabins ... if we had stayed in our original cabin ... the one we always used to stay in ... and if Deb hadn't refused to go inside ... and if we hadn't swapped cabins with Stephen ... it would have been me. I would have been standing on Stephen's floating platform ... and he would have been standing on ours ... and I would have been the one struck by the lightning ... not Stephen."

"That's a scary thought," says Mark.

"Yeah. I know."

"Deb, was that why you weren't able to go inside at the other cabin?" asks Mary. Because somehow, through this connection of yours, you knew about the lightning?"

"I don't know. I don't know what caused me to feel the way I did."

"But if it weren't for you Deb," says John, "I might be dead right now."

And then, there's another long period of silence.

It never crossed my mind that Deb's refusal to enter the other cabin might have saved John's life. "That's just faulty logic," I think to myself. Lightning strikes randomly. It could just as easily have hit our platform as Stephen's. Besides, if we had done something differently in the past, future events would have changed as well. If we had not changed cabins, everyone would have been in a different place, at a different time, doing different things, and John may never have jumped into the water. It's just silly for John to think that Deb saved him from the lightning. It's the same faulty logic people use after they have a car accident – where they reason that they could have avoided the accident if only they had traveled on a different route. They don't realize that changing your path also changes all the events you experience, and that the chances of having a similar accident on the new, alternate route you take, are just about the same as on the original.

"You know, I've often wondered," says John, in an almost dreamlike tone. "What happens to you when you die?"

"That's easy," replies Mary. "If you believe in Jesus, you go to heaven. Otherwise, you go to hell. It's pretty simple. It's what the Bible says."

"Actually," says Mark, "that's really not what the Bible says."

"Sure it does."

"No, I'm afraid it doesn't," continues Mark. "And people get confused about this all the time."

"How is that?" asks Matt.

"Well, the Bible talks a lot about events that happen far into the future – when there's judgment and there's a new Heaven and a new earth and there's no more pain or sickness or sorrow and everyone is happy and gets along with each other – but that could be thousands or millions or even billions of years from now. Way far into the future."

"But I thought you were judged right after you died," says John. "You know, when you meet Saint Peter at the pearly gates, and he decides whether or not to let you inside."

"I know that that's what a lot of people believe, but there's really no place in the Bible that says anything about Saint Peter manning a post at the heavenly gates."

"Really?"

"Yeah. Those are just stories that have been perpetuated by people that really don't have a very good understanding of the Bible – which is a category most Christians fall into."

"That's not true in my church," responds Mary.

"Maybe not," says Mark. But let me ask everyone a question. How many of you have read the Bible all the way through from cover to cover?" And when no one answers, he asks, "What about the New Testament?"

"I started to do that once," says Mary. "But I never quite finished. You know, I had work ... and my kid to take care of ... and ..."

"Yeah, we all get strapped for time, and so the only exposure most Christians get to the Bible is on Sunday mornings – when they hear passages read from the pulpit. And the result of that is ... that a majority of church goers are only exposed to a very small portion of scripture. You see, most Christians have never heard or read most of what is written in the Bible, and that leads to a multitude of misconceptions."

"And so, the Bible doesn't say that you go to Heaven when you die?" I ask.

"No."

"But Mark," says Mary, "almost all churches teach that if you believe in Jesus, you go to Heaven. You can't deny that."

"Yes, unfortunately, that's true. That's what they teach."

"So you think that they all are lying?" asks Mary.

"No, they're not really lying. But by taking small portions of scripture out of context, they obfuscate the whole truth."

"And so, what is the whole truth?"

"Well the truth is that we all may eventually get to Heaven, but it's not something that happens right away. First, you have to wait around a while for judgment day. There's no way to get into Heaven until you are judged – until you get a thumbs up from the almighty."

"And when exactly does that happen?"

"I don't know. But, according to Bible, before anyone is judged, Jesus has to return. And no one really has any idea when that will be.

"You mean to say that no one that has ever died has been judged? And no one has made it up to Heaven yet? Not even all those pre-historic cave men? Not even my grandma?"

"Yeah. That's really what the Bible says. No judgment until Jesus returns, and everyone who has already died is still waiting around for the big event."

"Brother David says that Jesus will probably return really soon," says Mary. "Within the next five years."

“You know,” replies Mark, “it's really quite funny how so many people can make so many bad predictions about when all the end-of-the-world stuff in the Bible is going to happen. For some reason, every generation thinks that Jesus' return is imminent, but throughout history, all the prognosticators have been wrong. It all started back in the first century, when Jesus' disciples thought that Jesus would return and judgment day would be within their lifetime. But it never came, and eventually, everyone died, and then the next generation waited anxiously for the second coming. And then the next generation, and the next, and on and on. And now, over two thousand years have passed – and still, no triumphal return and no judgment. And it's easy to envision nothing happening for another two thousand years ... or even much longer.”

“I don’t believe that,” says Mary. I could be raptured away at any minute, and then I'll be with Jesus in Heaven.”

“Well”, says Mark, “I doubt you will be. You see, just like the belief that you go directly to Heaven when you die ... the rapture is just another idea that is not really biblical.”

“No! That's not true!”

“Yes it is. And it goes back to the fact that people don't really know what the Bible actually says. But I can understand much of their confusion, because there are so many misinterpretations in biblical literature. So many, that it is often difficult to know what to believe – especially since many of these fallacies have been perpetuated by the church itself. And unfortunately, some of these have, over time, worked their way into mainstream Christian theology – like the rapture.”

“What's the rapture?” asks Deb.

“Well, the rapture is the belief, held by many denominations, that Jesus is going to return and whisk all the Christians off the face of the earth. One day everyone will be going about their normal daily routines, and then suddenly, all the believers vanish. And all the cars these faithful servants are driving at the time ... are suddenly driverless, and spin out of control and crash."

"I'd hate to be in an airplane at the time," says Deb.

"Me too," says Matt. "I bet all the non-believers left in the plane will be praying for the first time – hoping that one of the pilots is an agnostic."

"Probably so," says Mark. But all this turmoil is never really going to happen, because, in my humble opinion, the rapture is no more real than Saint Peter guarding the heavenly gates. It's not something the Bible really talks about. And even if you believe in a very literal interpretation of the Bible, where every verse in every book is connected with every other verse in every other book – and you piece together various out-of-context sections from all over – you still have to use a lot of imagination to come up with any good support for Christians being snatched away from the earth."

"Then why do so many people believe in the rapture?" asks John.

"Well," replies Mark, "I think it's because it makes for interesting sermons, and great press. The rapture was the main subject of all the 'Left Behind' books that were so popular a few years ago. And many of the people who read these books believed that they were based on biblical truths, and jumped on the rapture bandwagon. But I believe the rapture is really just a gross miss-interpretation of scripture."

"That's not what Brother David says."

"I'm sure it's not. But do you know how many times the word rapture is used in the entire Bible?"

"Well, since I can never seem to get any answers correct," says John, "I'll guess. How about ten?"

"Sorry John. Wrong again. The answer is actually zero. The Bible never mentions the word 'rapture'. And almost no one over the entire history of Christianity ever believed in the rapture until sometime in the mid nineteenth century when notes about a secret rapture we're added to a popular study Bible."

"But the rapture must have had some credibility," says John. "I mean, the people who write those notes are high powered biblical scholars, and I doubt that they would have added anything that was grossly misleading. Besides, aren't there lots of places in the book of Revelation that talk about Christians being taken away to Heaven?"

"No, Revelation doesn't mention anything about a rapture, or being left behind or flying away to meet Jesus. Those are mostly the ideas of a nineteenth century Jesuit priest, who took a few obscure, unrelated verses in the Bible, from books like Thessalonians, and formulated the theory of a secret rapture that has, for some unexplainable reason, been perpetuated in Christian theology."

"Revelation doesn't talk about the rapture?" asks Mary.

"No. Not at all."

"We'll," says Mary, "Brother David told us that the Bible says ... somewhere ... something like ... ah ... 'As in the days of Noah, two men shall be in the field, and one shall be taken and the other left behind.' So it may not be in Revelation, but it's in the Bible."

"Yeah, that passage is from the book of Matthew. But look at what it says, 'As in the days of Noah.' What happened in the days of Noah? Who was taken? It wasn't Noah and his family. They were safe on the ark. It was the wicked that were taken. The wicked were removed from the earth – not the righteous. And so it's really hard to use this verse for a justification that all Christians are going to fly away to heaven. It's more likely that the bad guys will be going away.

"You see, there are two problems with the way many people interpret scripture. The first is that they take various verses out of context, without any consideration of what is written before or after a certain passage. And the second is that they try to take out-of-context snippets of scripture and relate them to other out-of-context snippets – which leads to a gross misunderstanding of the original, intended meaning."

"So, what about that obscure verse in Thessalonians that you talked about?" asks John. "Have the rapture people taken that out of context too?

"Yes, they have. You see when Paul writes to the Thessalonians, he does tell them they will all meet Jesus in the sky. But if you read the whole chapter in context, you see that he's not describing a mass exodus from the earth. He's really just trying to reassure some insecure followers that everyone that has already died will not remain in their graves forever. He wants them to know that their deceased relatives will not miss out on salvation, but will be raised from the dead, just like everyone else. Paul just wants to reassure the Thessalonians that death is not the end to our existence – and that we will all, eventually, be reunited."

"But that just proves my point," says Mary. "We will all be raised from the dead and reunited with Jesus in Heaven. So, even if there is no rapture, which I doubt, we will all still be in heaven with Jesus."

"No. That's really not what the Bible says."

"Sure it is."

"No Mary. You see ... you're confusing the 'Heaven' thing and the 'eternal life' thing."

"What?"

"Heaven and eternal life are not synonyms. They aren't the same, and you really have to pay close attention to when the Bible uses each of these words, and be careful that you don't interchange their meanings."

"But they are really the same concept."

"No, they're not. You have to realize that eternity is a long time, and the first part of eternity may not be spent in Heaven. We may all want to believe that we go straight through the pearly gates when we die, but, again, that's not what the Bible teaches. The fact is that the Bible is very vague when it comes to saying anything about what happens to you immediately after you face the grim reaper."

"Are you sure about that?" asks John.

"Yeah, in fact, there are only two places in the Bible that even discuss what happens to you at the moment you die, and neither of those say anything about heaven."

"Really?"

"Yeah. So you see, if Stephen doesn't make it – if he were to die tonight in his hospital bed, there's very little to go on in the Bible to know what will happen to him tomorrow."

"I don't know how you can say that," says Mary. "The Bible says, 'For God so loved the world that he gave his only begotten Son, that whosoever believeth in him should not perish but have everlasting life.' So there!"

At this point, I'm glad I'm in my bed, and I don't have to participate in this conversation. It seems to me that arguments over the Bible are just a waste of time – especially when someone starts quoting that for-God-so-loved-the-world verse. There are so many contradictions in the Bible and so many different interpretations that no two people are ever going to agree on what any passage really means anyway. And why does it matter? In my opinion, this Bible, written by men thousands of years ago, has very little relevance today.

"Mary", says Mark. "You're interpreting that verse all wrong. And that's exactly why people get so confused.

That verse says 'should not perish but have everlasting life' – and for some reason, everyone assumes that 'everlasting life' means the same thing as 'Heaven'. But they are really two entirely different things. One's a state of being and the other's a place. And just because you have eternal life doesn't mean you're going to be spending the whole time in Heaven. That's the fallacy with the when-I-die-I-go-to-heaven logic. It's really not biblical, and that's what most people don't understand."

"But if you don't go to Heaven, at least not until after you're judged, whenever that may be, where do you go?" asks John.

"Again, the Bible really doesn't say much about the subject. Some Christians believe that you just fall asleep for a couple of eons until judgment day comes around. And then you have your day in court and receive your eternal sentence – to Heaven or to Hell."

"That doesn't sound very practical," says John. "I mean, just falling asleep for a few million years while waiting for a court date."

"Yeah, I know."

"At least you'll be well rested when you take the stand," says Deb.

"So, you think that, maybe, you go to someplace else in the meantime?" asks Matt.

"Yeah. That's another possibility that people over the ages have thought about."

"Like what?"

"Well, when the Bible was written in the first century, it was widely accepted by many people, including the Jews, that when anyone died they went to Hades."

"You mean everyone goes to Hell?" asks John.

"Oh no. Hades and Hell are two entirely different things. Hell is a place of punishment, but Hades is just a gathering place for dead people – where everyone goes to await judgment – both good people and bad people."

"But you said that judgment may be way far in the future. So, what do you do in the meantime in this Hades place? Play solitaire?"

"I prefer video poker myself," says Deb.

"I don't know exactly what the daily activities will be. But I'm hoping for something a little more intimate," replies Mark with a slight smile.

"I think you should be a little more worried about punishment than intimacy," says Mary.

"Yeah. Maybe. But unfortunately, we don't really know what will be on the agenda, because there's not a lot to go on. And because of that, most of what people believe about life after death is based on speculation and a lot of imagination. For instance, Catholics didn't like the long sleep or the Hades ideas, so they invented something called Purgatory. Now Purgatory is never mentioned in the Bible, but Catholics figured that since most people still had lots of faults when they died, they weren't really prepared to enter heaven. So, they reasoned that you must go to some intermediate place which they called Purgatory, where you could work on becoming a better person – so that you will be worthy of all the pleasures of Heaven."

"No!" says Mary. "You are saved by faith. Not by your works. Salvation is not something you earn by doing good things. It's a gift from God."

"True," replies Mark. "But what is salvation? What are you saved from?"

"You're saved from death."

"Yes. And so, again, you have eternal life – and again, that has nothing to do with going to Heaven. Eternal life and Heaven are two entirely different things. You have got to separate the two concepts – and then pay close attention to when the Bible talks about eternal life ... and when it talks about Heaven."

"So according to the Bible, how do you get to Heaven?" asks John.

"Well," replies Mark, "lets first talk about eternal life. The eternal life thing is easy. All you have to do is to believe in Jesus – or for those who lived in Old Testament times, you simply had to love God. And presto, you've nailed the eternal life thing. Not too difficult. You just accept all the abundant grace that God pours out on you, and you've got eternal life nailed."

"Then what about Heaven?" asks John.

"Well, getting to Heaven is a lot more difficult. Especially if you're rich. It's pretty easy for a rich guy to believe in Jesus and get his eternal life ticket punched, but Jesus said that it's harder for a camel to go through the eye of a needle than for that same guy to get into Heaven. I guess that's because you have to do a lot of very hard things. Like sell your Mercedes and give the money to the poor. And then you need to feed all the hungry people in India or wherever. And after that, you've got to deny yourself any of the worldly pleasures we love so much, and carry around a big burdensome cross. You see, getting into Heaven is a pretty hard thing to do. In fact, John the Baptist is even going to have a hard time getting there. Jesus says that the greatest man ever to walk on earth was John the Baptist, but then he goes on to say that the least in the kingdom of Heaven is greater than John."

"That's not true," says Mary. "Even the thief who was crucified on the cross next to Jesus made it. Jesus said to him, 'Today you shall be with me in paradise'."

"I guess he must not have been rich," says John.

"Probably not," says Mark.

"So," says Mary, "doesn't that prove my point? The thief dies and that very same day he's in Heaven."

"Well," says Mark, "Jesus's comment to the thief is the first place that the Bible talks about what happens to you after you die."

"So, I'm right," says Mary.

"No, not really," says Mark. "The Bible doesn't say the thief goes to 'Heaven'. It says 'paradise'. And that's something entirely different. And, if you go back and look at the original Greek language that the Bible was written in, the word that is translated as 'paradise' actually refers to a sort of holding area – a place where souls kinda wait around."

"Is that the same as that Hades place?"

"It might be."

"Could it be that spiritual world that Deb might have a connection to?"

"I don't know."

"What about the second place in the Bible that talks about where you go?" asks John.

"Well, the only other place in the Bible that talks about what happens to you immediately after you die is when Paul says that when he dies, he will be with Christ."

"Which is in Heaven," says Mary.

"Not necessarily. Jesus isn't under house arrest. I think he's probably allowed to roam around. In fact, Mary, if I were to ask you where Jesus is right now – what would you say."

"He's in my heart."

"Exactly. And I know you have a kind and generous heart – but it's not Heaven."

"So," says John, "Is that where we all go when we die? To that paradise, Hades, Purgatory place? And is that the place that Matt talked about that hides somewhere in all those extra dimensions?"

"I don't know."

"And is it also the place," continues John, "where time stands still, like at the center of a black hole. And is that where all the consciousness stuff emanates from – or to?"

"I don't know that either."

"So Matt. Can you help us out here?"

"Not really," says Matt. "I don't know either. No one does. Science just can't answer those questions."

"OK," says John. "Since Luke's not here to put a damper on the non-scientific speculation, let me ask you a question."

I'm still lying in my bed when I hear John mention my name. I have been meaning to get up for a while now, and go downstairs to join into this conversation – to be sure that no one says anything irrational. But the more that I listen, the more I begin to think that maybe ... I might not have all the answers. I long to believe, just like so many others, that there's something more to life – that your existence continues after you die, but I have never been able to justify it rationally. From everything I have seen and everything I have learned, the idea of life after death just doesn't make much sense. But maybe I've been wrong.

And maybe the science I have relied on for so long to formulate my opinions of how the world works has led me down the wrong path. A path that leads away from God, when a better understanding of the world – maybe an even a more advanced scientific understanding of the universe – might actually lead toward God. But that just seems kinda backwards. So, I'm confused – not sure what to think or believe. And so I just lie there. And listen.

"So, Matt, what do you think?" asks John. "I'm really curious. Is all this connected – everything we have been talking about the last couple of days? You know, with time stretching and the extra dimensions and all the consciousness stuff? Does it all fit together?"

"It's hard to say," replies Matt, but we do know some things. We know that time is not a constant, and that it can go faster and slower and can actually, under extreme circumstances, stop completely – and that makes the concept of eternity or of eternal life plausible – even for scientists."

"That was always a hard concept for me," says Mark. "You know, the eternity thing. It just never made any sense to me, but I never understood that time can stretch and shrink. And I never realized that Albert was onboard with all this."

"Me neither," says John.

"And then there's space," continues Matt. "Space that probably has six or seven dimensions that we can't see – dimensions that can possibly hide a multitude of things from us – like a parallel universe, or a spiritual world or even some heavenly realm. Or maybe all of those."

"And what about consciousness?" asks John.

"Well, that's probably the strangest thing of all, and something that is far beyond our current understanding. But we have observed small hints of how the world works on very small scales. And this has led us to believe that consciousness is a universal concept, and many intriguing studies have been performed that suggest that consciousness isn't just some electrical activity confined to our brain, but our consciousness actually interacts with the tiny particles that make up all of the matter around us."

"So then, you begin to wonder," says John. "If we were all made in the image of God ... and if somehow, consciousness can travel across the various dimensions of space all around us ... then perhaps our spirit, or our consciousness can somehow interact with God's. I know it's a real stretch, but isn't it possible, that perhaps, the mechanism that allows Jesus to be in Mary's heart – is somehow related to this consciousness thing?"

"Time to go downstairs," I think to myself. "John is beginning to draw unsubstantiated and irrational conclusions – trying to somehow justify God's presence through some quantum mechanical concept he really doesn't understand." And I am just about ready to get up when Matt intercedes.

"You're right," says Matt. "That is a stretch. And right now, we really don't understand enough about quantum mechanics to draw many conclusions – except that our world behaves in strange and mysterious ways – ways which are not evident to ordinary people."

Another long silent period ensues. Apparently, everyone is as confused as I am – just trying to make sense of things – questioning various firmly held beliefs that have become a bit unstable – wondering if the world is different than we had previously thought, and trying to integrate a few new pieces of information. We all seem to be wondering how our view of reality might be flawed, and we are all, in different ways, attempting to rationalize our beliefs about religion, the Bible and God. It's almost like Deb's puzzle – you don't know what the final result is going to be because there's no picture to go by, and so you just keep finding pieces that fit, until, slowly, the whole scene becomes a little clearer.

Before this week, I thought I understood all the puzzle pieces – and I had them assembled in a orderly manner into a nice, neat picture. It was so clean and uncomplicated. I took the Big Bang piece that explained the origins of the universe and put it together with the Darwin piece that explained the evolution of man, and added an experience piece that saw no evidence of God intervening in any worldly affairs. Those pieces fit nicely with the common sense piece that rationalized that science would eventually be able to explain the myriad of things we still do not understand. And I had a completed puzzle – with a picture of the world that fit together pretty well. It was a world that was created without a single spoken "Let there be" – a world that evolved without a master designer – and a world that runs without some predetermined plan – and a world without daily interventions of a supernatural power. And that all made sense to me.

But now, a couple of new puzzle pieces have been added – pieces that don't seem to fit very well into my current puzzle – my picture of how the world works. And somehow, I need to add these new pieces, but I'm not exactly sure where they are going to fit. How do I plug the extra dimensions of space into my neat uncluttered picture of the universe? And what does it really mean when time slows down – or even stops? And if consciousness really isn't confined to the axons and synapses in our heads, but actually travels around through time and space, maybe, just maybe, there is the possibility that our consciousness can live on after our physical death – even though that still sounds so absurd. And so, I just lie in bed, pondering a few new truths, theories and thoughts – weighing them against some very firm principles and beliefs that I have held on to for a long time. And I begin to grow weary and confused – not sure what to think, because now, the new, revised puzzle I am beginning to assemble is in transition, and my image of what the final picture will look like has ... well ... become a little foggy.

"Mark."
"Yeah."

“You know,” says Mary as she begins to cry, “I have always believed that someday, when I die, I will be reunited with Mom and Dad in Heaven. And now ... now you're trying to tell me that that isn't going to happen. And ... and ... I just can't ...”

“No, no, Mary,” says Mark as he walks over to Mary's chair and puts his arms around her. “I never said that. I believe we will all be reunited – with all of our loved ones.”

“Really?”

“Yeah. All I said was that we don't know a lot about what happens to us after we die. But I believe we do go to a better place. It's not a perfect place where there's no pain and suffering, and there probably won't be any clouds or harps – but it will be a better place. Probably some wonderful spiritual place. And although the Bible doesn't have a lot to say about this place, the Bible does have a lot to say about God's nature – that God is inherently relational, and that He desires us to be in a relationship with Him and with each other. And because of that, I believe that when we die, we will all be with the people that we love – in some way that we probably cannot even imagine. You can call it Hades or Purgatory if you like, but I just like to think that it's probably just a beautiful, spiritual place.”

“You really believe that's true?”

“Yeah,” says Mark, as Matt, Deb and John also come over to give Mary a hug.

I feel compelled to run down the stairs and put my arms around Mary as well. And my eyes begin to water as I imagine the possibility of being able to reconcile with my dad. To have the opportunity to finally tell him that I understand and accept him for who he was – or is. And to ask forgiveness for all the hurtful things that I have said and done. And to finally be able to tell him how much I care for him.

"If only I had a second chance," I think as a single tear runs down my cheek. "If only I could finally say, 'I love you'.

"But no, what am I thinking? How could that ever happen?"

"So you don't have to cry Mary, because Mom and Dad will be there waiting for us," says Mark. And after we die, we will all be together once again."

"You really think so?" asks Mary.

"Yeah," replies Mark. "And you know, I believe that someday Luke and his daughter will be reunited as well. And all the pain and agony that Luke has gone through over the years will be forgotten – replaced with an unimaginable joy. And they will experience a mutual love that is stronger than anything they have ever known before."

And when I hear Mark say that, I am overcome with sadness – and happiness. And I begin to cry – unable to control myself. I don't know why I am so overwhelmed with emotion, but it seems like all the tears I have suppressed for so many years have all been released – streaming down my face all at once. They are the tears from decades of silent sorrow, and from years of self-imposed solitude and endless mourning for the one person that made my life worth living. They are tears for that small, innocent child that was so mercilessly taken away, and tears for myself, a completely empty and broken man – the one who was left behind – the one who had to cope with a tragedy that no one should ever have to bare.

And as these tears continue to fall, I get up, and I walk out the bedroom door and down the hall, and I begin walking down the stairs.

"Luke," says Mark. "We didn't know you were up there."

“Are you OK?” asks John seeing my obvious distress. “Come on down with us.”

And all six of us gather together, and hold onto each other in front of the fire. Sobbing together. Happily. Feeling a bond – a family bond, holding us close together – unlike anything this family has ever experienced.

CHAPTER 5 – DAY 4

Only two things are infinite, the universe and human stupidity, and I'm not sure about the former.
Albert Einstein

Luke's Problem

I never usually eat breakfast at home, but when I do, I normally just have a piece of fruit, or some juice or perhaps a small bowl of oatmeal. But for some reason, the breakfast at Deephaven compels me to lose any motivation for maintaining my standard low calorie, low fat and low cholesterol regimen, and somehow, induces me to imbibe in cuisine that I usually consider taboo.

So, every morning, I stack my plate with fried eggs, bacon, and a whole handful of those little sausage links that I love so much. And after finishing those, I usually go back to the buffet for a Belgium waffle that I smother with whipped butter and maple syrup – syrup which is boiled to sweet perfection just a few miles from the camp.

I don't know why I act this way, especially when there is such a wide assortment of healthy fruits and cereals. Perhaps it's the hypnotic aroma of the fresh brewed coffee that subconsciously lures me toward the buffet line. Or perhaps it's just that my appetite is enhanced by the crisp, clear air that periodically causes you to take in a deep breath, to savor the fresh scent of air that lacks ozone, sulfur dioxide and various carcinogenic particulates.

And just as I take one of those extended breaths, and as I twirl my last piece of waffle across my plate – trying to soak up the last bit of syrup, Deb says to me, “Luke, it's a beautiful morning. Why don't you walk back to the cabin with me. We can take the long route, you know, along the lake.”

“Sure,” I say. “I probably need to walk off a little of this breakfast anyway.”

Deb and I get up from our familiar, green painted dining table with the rustic white birch legs, and walk past the buffet where I grab a biscuit to go while Deb refills her coffee cup. I hold the biscuit in my mouth as I use both hands open the heavy dining hall door that leads to the porch – the long covered porch with the wavy wooden planks that have been worn thin by generations of families coming to the dining hall for their daily meals.

We then walk down the creaky wooden steps and past the library – the library with those old wooden shelves that lure you inside with that distinctive musty smell of old classics – books with brown edges and torn bindings – books written by Mark Twain and Ernest Hemingway and Stephen King – all of them waiting patiently to be chosen for another relaxing afternoon under a tree by the lake – and perhaps, hoping, after a few chapters, for a brief snooze in the shade.

Next to the library is the camp office with the friendly staff that will drop everything they are doing just to chat. Any topic is fine – anything you might feel like talking about. We wave to the two young college kids manning the front desk and they smile and wave back. Genuine smiles. Smiles that you don't often see in department stores or fast food restaurants.

Then we follow the dirt and rock path toward the lake that passes close to the kid’s playhouse and near to the area where rowing sculls are stored.

When we get close to the shore, I pick up a flat rounded stone from the path, and I toss the rock sidearm toward the center of the lake. The rock skims the top of the water – hitting the surface eight, ten, twelve times, before finally sinking to the bottom.

"Good throw," says Deb. "Let me try." Deb picks up anther stone and attempts a similar sidearm throw, but the rock just plunks once, and disappears below the surface. "Guess I need some practice."

"Yeah. You need to find a flatter stone, and you need to be sure it's spinning when it leaves you hand. Go ahead. Try again."

"No. I don't think I'm very good at this."

"Ah, come on Deb. You can do it. Come on. What are you afraid of?"

"You know Luke, that's what I wanted to talk to you about. I think you may be the one that's afraid."

And that's when I realize what the purpose of our little excursion is, and why Deb asked me to walk back to the cabin with her. It was just an excuse to trap me into an informal therapy session, at a time when my defenses are down and I am the most vulnerable. It is a sly tactic on her part, but for some reason, I don't take issue with her divisiveness. In fact, it's almost a welcome relief, because I know that I have issues that I need to talk through, and now may be as good a time as any to get a professional opinion of my deep seeded fears and my unexplained paralysis.

"So," I say, "I guess you noticed that I wasn't around yesterday when the lightning hit Stephen."

"Yeah."

"Do you know where I was?"

"Yeah. I heard you in the rocking chair upstairs when I came inside for a few minutes to work on the puzzle."

"Was that before you ran outside to warn everyone about the storm?"

"Yeah."

"So you knew I was upstairs the whole time?"

"Yeah."

Deb takes a seat on one of the large granite boulders along the shore and motions to me to sit down as well. The sun is just beginning to rise above the mountains on the opposite side of the lake and Deb considerately positions me in the shade, probably thinking that any bright light in my eyes might feel too much like an interrogation. And then she continues very cautiously, “Do you think you might want to talk about it?”

“I guess so. But somehow, I think you may already know what my problem is.”

“You know, I'm not really sure, but that's unimportant. The real question is, do you know what the problem is?”

“No, I'm afraid I don't. I know what some of my symptoms are, but I'm not sure I understand the underlying cause. I don't know why I react like I do.”

“Tell me about the symptoms.”

“Well, it's really just an overwhelming and paralyzing fear – in any life-threatening situation.”

“And when did you first notice this?”

“I guess it was right after my daughter died.”

“After she was hit by the car?”

“Yeah. She was riding her tricycle, and I thought my wife was watching her, and she obviously thought I was.” And then I pause, and swallow hard, and try to hold back some of the pain I am feeling. And I close my eyes briefly, and then slowly continue. “And then ... she rode out ... out into the street ... out in-between two parked cars ... out into oncoming traffic ... and out of my life ... forever.”

“There's nothing in this world that is as painful as losing a child.”

“Yeah,” I say as I look down toward the ground. “But the most painful part was six days later in the hospital, when they took her off life support. That's when my world and my life crumbled. That's when I experienced the real agony of death. That's when I first felt that debilitating fear.

The fear that comes from knowing that someone you love so much ... is gone ... and is never coming back."

"How long did that feeling last?"

"It never really went away. And starting from the very moment they pulled the plug on the ventilator, my life went into a downward spiral, and it wasn't until many years later that I was even able to begin dealing with the grief and the loss I was feeling."

"What about your wife?"

"She was devastated as well, and she desperately needed me to help her cope with our loss. But I was worthless, and my wife had to bear the burden of my grief as well as her own. And within two months of the funeral she became so frustrated from the lack of any love or encouragement or support from me, that she moved out of the house and filed for divorce.

"So, the entire course of my life was changed in just a matter of seconds – after just a brief moment of neglect – the time it takes for the wheels on a small tricycle to spin around just three or four times. And in that indescribably short period of time, a precious life ... just ... slipped away. Gone forever. And ever since then, whenever I am faced with a life threatening situation, I relive the moment I heard the car's brakes screech ... and I saw the smashed tricycle ... and I stood over my daughter's battered body ... lying ... practically lifeless ... in the street. And then, I re-experience the helplessness and fear I had in that hospital room when we were told that there was no hope, and I feel the pain from the wounds in my soul that have never really healed ... and I am overwhelmed both physically and emotionally ... and I am unable to respond or react normally."

"I think I understand. And I think you understand too."

"Yeah. But this is the first time I have really been able to express it to anyone."

"Yeah."

"It was just a few months after my wife filed for divorce that I flunked out of medical school at the University of Illinois, and from that point on, I began spending the majority of my time just lying on my couch – longing for the life I once had, before it was so mercilessly taken away from me."

"So that's when you retreated from the world."

"Yeah. For a number of years I really had no social contacts. I was just in too much pain."

"And now?"

"Now? They say that time heals all wounds, and it has to some degree. But the scars are still there. And I don't know if they will ever completely heal. Nan has helped me tremendously, but there is only so much she can do, and I'm afraid that lately, I have been a bit of a disappointment to her. And I'm afraid that our relationship is suffering because of my inability to move on ... to live life without the shackles of the past ... and to be able to give Nan the love and support that she deserves – without the baggage of memories that haunt me daily ... or the nightmares that interrupt my sleep ... painful experiences that leave me longing to be by myself ... alone in my misery as I try to cope with a deep-seated injury that never seems to go away."

"Maybe you're trying to do too much on your own. Maybe you need to let others help you a little more."

"Yeah. I know. But it's hard for me."

"Yeah, but you are beginning to talk about it now. And I know it's difficult to discuss these things, but think that our talk this morning is a big step forward."

"You think so?"

"Yeah. We'll talk more later. Right now, we probably need to get back to the cabin. We have a puzzle to work on."

"OK."

As we walk back to the cabin along the rocky path around the lake we don't say much more to each other, but I feel a sense a warmth and camaraderie as we stroll along side by side – which comforts me and gives me hope – hope that there may be a way to work through some of my recurrent problems, and hope that I can build a better life for Nan and for myself.

It is not until about halfway back to the cabin I realize that I didn't mention anything to Deb about my cancer returning – and I realize that that subject probably would have been pertinent to our conversation – with death now staring me directly in the face. "But it may be better to wait before I broach that subject," I think to myself. "At least until after the wedding."

When Deb and I arrive back at the cabin, we walk in to find Matt, Mark, John and Mary all huddled around the puzzle. In the background there is music playing, and my four siblings are singing and swaying back and forth to the beat – while, at the same time, scanning for missing puzzle pieces. I turn to see where the music is coming from and I notice a rocking chair near the fire where someone has hooked up their cell phone to a small portable boom box to play their music library.

I recognize the song playing immediately – it's a Beatles' classic written by John Lennon called "Come Together", which was released on the Abby Road album in 1969. In my opinion, one of the greatest albums of all time. It was the last album the group recorded in the studio before the band broke up.

After listening for a moment to the quartet around the puzzle, I realize that, apparently, there is very little musical talent in my entire family. Everyone seems to be singing the song in a different key, but I guess it doesn't matter – so I join in as well.

Come together right now over me.

It would be hard for anyone to recognize the melody amid the discord, but the words are unmistakable. "And how appropriate," I think to myself. "Come together right now." Maybe it's an omen – a sign that this family can come together ... now ... right now. Maybe despite all our differences, we can make this family work. Together.

By 1969 all the Beatles had gone their separate ways, and no one was getting along with any of the other band members. And it's generally accepted that the verses of this song talk about the conflicts created by each of the group's members. And if you look at the lyrics, and think of them in this context, the words, which at first seem like gibberish, begin to make sense.

He wear no shoe-shine
He got toe-jam football

Lead guitarist George Harrison had become enamored with the Indian Maharishi Mahesh Yogi, and would often wear simple robes and bare feet (no shoe shine), and play bare foot soccer at the yogi's temple (toe-jam football).

He got Ono sideboard
He one spinal cracker

John Lennon's girlfriend Yoko Ono (Ono sideboard) became a big distraction for the rest of the group, and John's constant complaining was breaking the backs of the other band members (spinal cracker).

He say one and one and one is three
Got to be good looking 'cause he's so hard to see

Paul McCartney, the 'good looking' Beatle was spending more and more time away from the band (hard to see), and had been talking about leaving the other three in the group (one and one and one is three).

Got to be a joker he just do what he please

And finally, Ringo Starr, the often criticized drummer, was the cut-up of the group, and often did bizarre and unexpected things.

So there they were, four talented and bright musicians – each gone their separate way. Not much in common anymore. All of them ready to cut the bonds – to break up the band and go off on their own.

But they came together and created an album that soared off the charts. One that has gone down in the annals of music history. And so, I think again to myself. "This is an omen." This song is telling us that we can make it. This family can come together. Despite all our differences. Right now. And so we all sing ... together.

Oh come together
Yeah come together
Yeah come together
Yeah come together

And in unison, as we bellow out these repetitive lyrics, we work on the puzzle, and laugh, and sway to the right and then to the left – together. It's as if the divisions between us have evaporated into thin air – or, perhaps, it's just that the differences between us were misconceptions – exaggerated childhood memories that had invaded our subconscious minds, that lingered and grew over our years of separation.

So maybe, our divisions are just imaginary, existing only in vague and distorted images from our past. Perhaps all we really ever needed to repair our broken relationships ... was to gather together once again, in one place, and rediscover the things we have in common. Yeah, rediscover our family bonds. Yeah, to come together.

Yeah come together
Yeah come together
Yeah come together

“Yeah,” I think to myself. “We are all coming together. Right now”

But then again ... I remember ... right after the Beatles came together ... and put aside their differences ... and recorded Abby Road ... probably their finest work ...

They broke up – and The Beatles never played together again.

"So, is everyone having fun working on my puzzle?" asks Deb.

"No!" responds Mark, still swaying in time with the music. "I hate this thing. It's way too hard."

"It's not that bad," says Mary. "The cabin in the center is starting to come together, but all the pieces look alike, and it's hard to find any that go together. It sure would help to have a picture to go by."

"Sorry."

"You know," says John, "I have been trying to find this last border piece for almost an hour. It should be easy – because it has a flat edge. But it's just not here. Nowhere. It's as if it were swallowed by a black hole."

"Well," says Matt, "if it was swallowed by a black hole, you will never see it again. Nothing escapes from a black hole."

"Oh no," says Deb. "John, you should never have used the 'B' word."

"So Matt, since we don't have anything else to do until lunch, can you explain a little more to us about these holes that are all black?" asks John. "I'm not sure I really understand much about them. All I really know is that they have a lot of gravity."

"Maybe I'll take a 25 or 30 mile walk so I don't have to listen to any of this," says Deb.

"Oh, this won't take long," says Matt. "Just five or ten minutes."

"Maybe I'll just go upstairs and pluck my eyebrows."

"Why don't you just work on the puzzle, and try not to listen."

"OK", replies Deb as she puts her fingers in her ears.

"Alright Matt," says Mark, "tell us about black holes."

"Wait just a minute," interrupts John as he runs over to the cell phone in the rocking chair next to the fire. "I have to put on some appropriate music."

Of course, we all know what he is going to play before he even picks up the phone. It was obvious to everyone. And after a few clicks on the screen, John and Mick Jagger start singing a classic Rolling Stones hit from 1966 – "Paint it Black".

I see a red door and I want it painted black
No colors anymore I want them to turn black

"OK Matt," says John. "Now that I've set the mood, explain to me how this black hole beast devoured my puzzle piece."

"Well," replies Matt, "a black hole can devour just about anything, big or small."

I see a line of cars and they're all painted black

"In fact, anything that gets close to a black hole is pretty much history – never to be seen again."

With flowers and my love both never to come back

"But in order to really understand what a black hole is, you have to understand how one is formed. So we'll start at the beginning. Now John ..."

"Yeah."

"We know the sun is pretty big. How much more mass do you think the sun has than the earth?"

"Oh, I don't know. But I would guess that the sun has 1,000 or maybe even 10,000 times the mass of the earth. Is that close?"

"No. Not even in the ballpark. The sun's mass is 330,000 times larger than earth's. It's really big, and really heavy, especially compared to the puny little planet we live on."

"I guess it has to be really massive to keep all the planets in orbit."

"Yes, but that large gravitational field also serves another purpose. It holds the sun together. You see, all the nuclear fusion reactions going on inside a sun produce an outward force, pushing all the matter in the star out into space – just like any nuclear explosion creates a shock wave emanating outward. But at the same time, the outward forces are counterbalanced by the sun's tremendous gravity – gravity that pulls everything back in. And eventually, an equilibrium is reached, where the outward and the inward forces balance. And so, the sun's outer edge becomes stable, and we see our sun with its constant current diameter of about 865,000 miles across."

"So what does that have to do with black holes?"

"Well, what happens when a star runs out of gas?"

"I guess there are no gas stations in space."

"No. And when a star's fuel is exhausted, it can no longer support its nuclear fusion reactions, and so the outward force created by these explosions fizzle, and the inward force, gravity, takes over."

"And so the sun implodes inward and turns into a black hole?"

I wanna see the sun blotted out from the sky
I wanna see it painted, painted, painted, painted black

"No, it won't. Our sun will never be able to form a black hole. It will die a relatively quiet death. It's not big enough. Even though it has 330,000 times the mass of the earth, it still doesn't have enough mass to generate a black hole. But it's a different story for larger stars. When big stars with the mass of 8 or 10 or 20 or 30 times the mass of the sun run out of fuel some catastrophic things begin to happen."

"There are stars that have 30 times the mass of the sun?"

"Yeah, lots of them. And when these stars run out of gas, their tremendous gravity causes their inner iron core to collapse – and all the heat and pressure and friction and tidal forces from this inward flow causes the star to explode. A giant explosion. A supernova. One of the largest cataclysmic events we know of in the universe."

"Like the planet Krypton exploding? Or like that planet that got in the way of that Death Star?"

"No, those are just small planets. Imaginary planets at that. What we are talking about here is something much, much larger and incredibly more powerful."

"So the star explodes and disintegrates?" asks Mary.

"No. The core of the star remains, because of the tremendous gravity. It's just the outer layers of the star are ejected into space by the supernova."

"So ... what happens to the rest – the stuff that's left behind?"

"Well, it depends upon how much mass is left in the star after the explosion."

"So?"

"Well, if the total mass left in the star is between 1.4 and 3 times the sun's mass, the star will collapse inward and form a neutron star."

"Neutron star? I thought we were talking about black holes."

"Just wait a few minutes. We're getting there."

"OK, so, what's a neutron star?"

"It's a very small but very dense star remnant. They are typically only 6 to 10 miles in diameter but they contain the same mass as one and a half to three suns."

"In a star only 6 miles in diameter?"

"Yeah. You see, the stuff left behind after a supernova explosion is very compact. In fact, one teaspoon of material from a neutron star would weigh about a billion tons on earth."

"What?"

"How could that be?"

"Well, it's because the tremendous gravity squeezes out all the empty space between the atoms, and you are left with something that is really heavy."

"But a spoonful weighs a billion tons?"

"Yeah, you see, in most materials that we are familiar with, the atoms and molecules that make up matter are separated from each other by great distances. Lots of empty space. But in a neutron star, all that space is squeezed out, and you end up with a bunch of neutrons stacked right on top of each other. And even though these neutrons are really small, they are really heavy when there is no empty space between them."

"But if all the neutrons end up stacked on top of each other, what happens to the protons and electrons and other stuff that is in a star?" asks John.

"Good question. You see, when the immense gravity of these stars forces out the empty space between atoms, all the electrons that once orbited the atoms are squeezed into the atom's nucleus, and these electrons combine with the protons, which changes them into neutrons – a positive proton combining with a negative electron to form a neutral neutron. And you end up with a clump of matter composed almost entirely of neutrons stacked tightly together. And it's really heavy."

"But what does that have to do with black holes?"

"Well, now that you know that dying stars up to three times the mass of the sun can collapse into neutron stars, what do you think happens to dying stars that have seven or ten or twenty or thirty times the mass of the sun?"

"Ah, a black hole?" asks John.

"Yes. You see, when you have that much mass, gravity becomes unimaginably strong. And after gravity squeezes out all the empty space between the atoms in this star remnant, and after there are only neutrons left, just like in a neutron star ..."

"Yeah?"

"Well, gravity in these stars is so strong that it crushes the neutrons into oblivion. All of them. And it crushes the quarks that make up the neutrons – squashes them completely. And even though the total mass of our star remains the same, this mass begins to occupy a smaller and smaller area. And this shrinking process continues, until finally, and I know this is very difficult to conceptualize, but finally, gravity crushes the entire mass of that giant star into an indescribably tiny point."

"Like, as small as the size of a pin head?"

"Oh no. Much smaller than that."

"Really?"

"Yeah. And when you have that much mass in that small a space, incredible things happen. Gravity is so strong that anything that gets close to the black hole is sucked into it, and is crushed into that tiny center point – just like everything else."

"Even our missing puzzle piece?"

"Yeah. And remember how we said that light was made up of small particles, and that these particles have mass?"

"Yeah."

"Well, if light gets within a certain distance from a black hole, it gets caught in its gravitational field, and it gets sucked in as well. Not even light can escape its tremendous gravity."

"So, how close can light get before it gets trapped?"

"Well, that depends upon how much mass the black hole has. For small black holes formed by the collapse of a star, light might have to get within 10 or 20 miles before it gets trapped. But black holes can become extremely massive as they gobble up more and more material, and eventually they can contain millions of times more mass than the sun. And in that case, light may be trapped within a much larger radius – sometimes millions of miles wide."

"Wow."

"So what happens to all this stuff when it gets sucked in and crushed" asks John. "Doesn't it have to go somewhere or do something? I mean, I have trouble understanding how everything gets smashed into a tiny little point. What's really going on inside this black monster?"

"Well, we really don't know the answer to that. Light can't escape from a wide area around a black hole, so there's no way we will ever be able to see beyond that horizon. But even if we could, we still probably would not be able to understand what is going on. You see, all the current theories we have ... all the theories that explain how the universe functions ... break down in black holes. They don't work."

"Hugh?"

"Yeah. Albert's theory of general relativity that explains how gravity works – just falls apart in very small spaces. And the theories that explain small spaces, quantum mechanics, don't work when there is a huge amount of gravity. So, we really understand very little about what happens inside one of these beasts."

"But can't scientists guess, or make estimates about what is going on inside?"

“Yeah, they can. And they do. But the things they come up with sound more like science fiction than science.”

“Like what?”

“Like wormholes to other parts of the universe – or to entirely different universes. Or linkages to other dimensions of space and time. Or ...”

And then there's a knock at the front door.

"Excuse me. Is anyone home?"

We all turn toward the front door and see Peter with a load of firewood in his arms.

"I brought you some extra firewood." says Peter. "It looks like a front may be coming through this afternoon, and it may get pretty cold tonight."

"Great."

"Peter," says Deb, obviously trying to change the conversation away from black holes to something a little more relevant, "Have you heard anything about Stephen? Is he still in the hospital? Is he going to be OK?"

"Well, we just received an update from his family this morning, and I'm afraid the latest news about Stephen is not good. He's still in a coma. He hasn't regained consciousness since the accident."

"That's so sad."

"Yeah, and the doctors are not real hopeful," says Peter as he stacks the firewood in the corner. "Stephen's heart is very weak, and his brain function, as best they can tell, is very marginal."

"Oh no."

"Yeah"

"What about his family?"

"Well, they are all staying at a hotel over near the hospital."

"They must be devastated."

"Yeah, they are. But I have known them for a long time. And they will pull through this. They have a very strong faith."

"Faith can only get you so far," I say. "I know what it's like ... waiting ... hoping for a miracle ... thinking that any moment God is going to intervene ... that He is going to pull you through ... that everything is going to be OK ... and that life is going to return to normal."

"There is always that hope. It's what we have been praying for."

"Yeah, but it doesn't work that way. God doesn't intervene. He just watches. And He teases you by offering you a false sense of hope, and then He just sits back and watches as lives fall apart, and everyone around you suffers – and you suffer. God really doesn't do anything to help in your greatest moments of need. And sometimes, I even think He delights in the pain that we go through while we wait helplessly and hopelessly, watching someone you love die."

"You know," says Peter as he takes a seat in the rocking chair next to the fireplace, "I know that death is very difficult to deal with. And I know how hard it is to lose a loved one. And I know that Stephen's family is in great anguish right now. And so, I understand why you feel the way you do."

"So Peter, have you lost someone close to you?" asks Deb.

"Yeah, my little sister died four years ago. She had a rare, untreatable cerebral tumor."

"I'm so sorry."

"It's OK. It was hard for the whole family at first, but we believe that there is more to life after we die. We believe that our spirits continue on. And we were all comforted by that – knowing that my little sister has passed through death, and is now living in a better place."

"How can you be so sure?" I ask.

"Well, it's hard to explain. It's just something you know. Something you feel deep inside."

"But weren't you angry? That she died so young? That God took her away when she was just a child?"

"The grief we all felt is hard to put into words. But there was very little anger. You see, our family has a strong faith in God. And that faith is what sustained us through those difficult times. That's not to say that accepting my sister's death was easy. It wasn't. Still isn't. But I believe that we all have a purpose here on earth, and we are all placed here for a certain period of time to fulfill that purpose. And so, even though we don't understand why she was taken away from us, our family feels that we were all blessed to have had Sis with us, and part of our lives – for over 12 years."

"That's one way to look at it," I reply. "But another viewpoint is that your sister's best years were taken away from her – that she was never able to experience things like her high school graduation, or her first love, or getting married and having children of her own. And no matter how blessed you believe you were to have her around through her early teens, the fact remains that she was robbed of a full life – a life that could have been so meaningful for her and all the people that she loved – and the people who loved her."

"Yeah, that's true. But I just have faith that the experiences she is having now are just as fulfilling, and perhaps even more rewarding than anything she would have been able to experience here on earth."

"I sometimes wish I had a faith like that," I say, half-jokingly.

But then Peter turns and looks straight at me and says in a pleasant but very serious voice, "I think someday you will. In fact, I know you will. I have faith in that."

I am not sure what Peter means by that comment, and I pause for a moment, trying to think of how to respond – searching for just the right words to express the way I feel – to explain that faith is just a crutch that people use to attribute random acts of nature to God – to build hope in times of crisis – to abdicate responsibility for your own actions to some higher deity. But as I begin to formulate my thoughts, I start to stutter, and I have trouble forming my words, and I lose my concentration, and all I am able to say is, "Y-Y-Y-Yeah."

"Well, I have to go," says Peter as he gets up from his rocking chair. "Lots more firewood and ice to deliver. See you guys later."

"Bye Peter,"

"Bye."

But then Peter stops for a moment as he's walking out the front door and turns and says, "Oh by the way Luke, I almost forgot."

"Yeah."

"Some doctor has been trying to get hold of you. He keeps calling the front office. Dr. Plank I think. He says it's important, but he won't leave a message, because of something about HIPPO or HIPPA – whatever that is. But, anyway, he wants to talk to you."

"Thanks Peter. I know what it's about."

"OK. Bye."

"Isn't Doctor Plank your oncologist?" asks Deb.

"Yeah."

"You better call him."

"I'd let you use my cell phone but I haven't been able to get any service in the cabin lately," says John as he picks up his phone off the rocking chair. "So Luke, why don't you go up to the office and you can call your doctor on the land line."

"Oh, that's OK. It's probably not necessary right now."

"But aren't you supposed to get the results from your cancer scans?"

"Yeah, but it can wait."

"But Luke, this doctor is obviously trying to get hold of you. It might be important. Don't you want to give him a call?"

"Not right now."

"How come?"

"Well, I already talked to him briefly," I say, hoping to end the conversation.

But Deb is not deterred, and she pursues the discussion, "So, what did he say?"

"Oh nothing important."

"Luke."

"Ah, well, I just talked to him for a minute, but then my cellphone died. No service."

"And what did he say?" And after I don't respond for a moment, Deb repeats, "Luke? What did he say?"

"He just wants to run some more tests."

"Luke," says Deb, becoming a little impatient. "Tell us what's going on."

"They're not sure. They just need to do some more testing."

"Has you cancer come back? ... Luke? ... Luke, you can talk to us. We're family. You don't have to face this alone. You are going to need the help and support of others. You need to talk about this."

"I know," I say, not looking up from the floor. "It's just hard for me."

"What did the doctor say?"

"Ah ... well ... he said ... that it looks like there are a couple of ... ah ... new tumors."

"Oh Luke, I'm so sorry."

"Yeah."

"Does Nan know?" asks Mark.

"No, and you have to promise not to tell her – at least not until after the wedding. Today's an important day for her, and I don't want any distractions."

"I think we all understand," says John as everyone else nods in agreement.

"You know Luke," says Matt, "you don't have to fight this thing alone. You have your family. You have all of us. And we can help you through this."

"That's right," adds Matt. "We'll be right at your side. And we'll help you beat this thing back into remission."

"Yeah, just like you did before," adds John, "but this time, we'll be there with you, and we'll fight it together."

"Thanks guys. I appreciate that. You all are great. I am so glad we were able to get together this week ... in this place," I say as we all gather around the puzzle.

"And you know," says Mary, "I think everything will work out fine. I have faith you will be healed. I believe that in my heart."

"Yeah," I say, "but you have to realize that when these cancers come back, especially after seven years, they are usually very aggressive. And the prognosis is not great."

"But they are bound to have new drugs and new treatments now. I mean, they come out with new things every year. Don't they?"

"Yeah they do. But as far as I know, there have not been any breakthroughs with this type of cancer."

"But you beat it once before."

"Yeah, but round two is always a lot more difficult, and survival rates are not nearly as good."

"But I am sure you will have many good years ahead of you – fulfilling years – years with your family at your side – a family that will help you through some of the difficult times ahead," says Deb.

"And you know," says Matt, "all that time you spent in the past – by yourself – fighting depression and fighting your cancer – all that time you spent in isolation – that time will seem insignificant compared to the time we will have together."

"You think so?"

"Yeah, I'm sure of it."

"So Matt," I say. "Let me ask you a few questions."

"Sure."

"I have been thinking a lot about time lately. You know, how much more time I'm going to have to spend in chemotherapy – how much time I have before I have to face the dreaded hospice social worker – how much time I have left before I have to say good-bye to Nan. I guess you do that when you finally realize that your life will eventually end."

"We are all going to die," says Matt. "And we will all have to face those things at some time."

"Yeah, but I'm still a little confused about time – what it really is – and what it really means. And I'm confused about what eternity is. I mean, how can a rational person believe in time that doesn't end or time that stands still? And how can a person well-grounded in the natural sciences believe that someone's spirit can go on living forever? How can you believe in eternal life – when life seems so finite? It's just hard for me to understand. It doesn't make sense. So Matt, I need to learn more about time. I know that time stretches and shrinks, but I need to understand how time stops inside a black hole. How does all that work?"

"Well, you have to realize that our limited view of the world is flawed. The universe does not function according to our common sense perceptions and ideas. And when you begin to understand some of the laws that govern the way things really work, you are entering a very foreign and mysterious realm."

"So how does time function in this strange universe of ours?"

"Well, do you remember why time slows down when you are traveling really fast?"

"Yeah. It's because light has a speed limit. And even though two people may be traveling three quarters the speed of light toward each other – it's impossible for an observer to see them approach each other at one and a half times the speed of light, because nothing can go faster than the speed of light. So, time slows down."

"Good."

"But I don't remember why time slows down in a strong gravitational field."

"Well, once you understand that time slows with speed, the gravitation slowing of time is pretty easy to comprehend. You see, Albert realized that acceleration and gravity are really the same thing. A person in an accelerating elevator in outer space may think that he is actually standing on earth, and a person standing on earth may actually believe he is in an accelerating elevator in outer space. You see, if this person's eyes are closed, he would have no idea which one of the two places he is standing. To him, there's really no difference. Gravity and acceleration have exactly the same effect."

"That's right. I remember."

"And if we observe a person falling inward toward a black hole, we would see this person's watch tick slower and slower compared to ours. Until finally, when our intrepid space explorer crosses the boundary where even light cannot escape a black hole, we see his watch stop completely."

"So there is no time."

"Well that's really hard to say. We don't really know what's going on inside a black hole, but what we can say is that relative to our time, time inside a black hole stands still."

"Yeah," I say. "And I have been trying to comprehend exactly what all that means. I mean, is that what eternity is? You know, when time goes on forever – or maybe when time stops. Is that how God can live outside the bounds of our time?"

"I don't know."

"So, is there, perhaps, a scientific explanation for life after death – everlasting life? I mean, can an intelligent, rational person somehow justify a belief that you can go on living forever?"

"Well," says Matt, "no one could survive falling into a black hole. Your body would just be torn apart. Literally limb from limb."

"What about someone's spirit?" asks Deb. "Could it survive?"

"I didn't think you were interested in this conversation, Deb" says Matt.

"Just answer the question Honey."

"Well, the answer is, I don't know. Because I don't know what a spirit is. I've never seen one."

“Well then, if maybe, someone's spirit or consciousness can somehow travel outside their body – like we talked about before – could it perhaps, even travel through a black hole – to a place where time stands still?”

“That's another question that is impossible to answer. No one has ever seen the inside of a black hole, and no one has ever seen a consciousness. So we really don't know. We can speculate all day long, and we can imagine all sorts of strange and wonderful things that can go on inside the gravitational abyss of a black hole. But there is no way we can test or verify our theories. We just don't have any good answers.”

“Got it!” shouts Deb.

“You have a good answer?”

“Oh no. I got 'IT'. That last border puzzle piece we have all been looking for. The one with the flat edge. It wasn't swallowed by a black hole after all. It was right here in front of our eyes all the time.”

Fishing

"Look what I have," says John as he walks into the cabin with the screen door slamming behind him.

"Fishing poles?" replies Deb.

"Yeah. I bought them at the camp store. Two of them."

"I'm not sure you can call a two and a half foot long stick with a Disney character reel, a 'fishing pole'. These things are not really made for fishing," I say as I examine the contents of one of the poles in its cardboard and plastic packaging. "They are just meant to keep small children occupied while their parents lounge in the sun."

"Nonsense. You've got everything you need here to catch a fish. A rod, a reel, a sinker, a float, 8 pound test monofilament, and a hook. And you get all that for only sixteen dollars!"

"What a deal."

"Yeah, and it even includes a Snow White tackle box."

"Great."

"Come on! Let's go fishing. I also got worms."

"Is that what smells so bad?"

"Yeah, but I bet that five pound small-mouth bass at the end of the dock can't wait to get his jaws around one of these delectable, slimy creatures."

So, since none of us want to circumvent John's attempt to arrange a family fishing adventure, we all acquiesce and dutifully march down to the dock. Besides, it's a beautiful morning and spending an hour or so on the water before lunch may not be such a bad idea. And I remember something Dad said once ... on this very same lake so long ago ... something that I didn't understand ... something, at the time that I thought was dumb. He said, "Luke, fishing has nothing to do with catching fish." And as that memory runs through my mind, I begin to think that maybe my Dad might have been a little smarter than I had given him credit for so many years ago.

"Do you remember how to tie a fishing knot?" asks Deb as she rips open one of the packages.

"Sure," says Matt. "Thread the line through the hook. Then wrap it around three or four times, and push the end through the bottom loop, and then through the top loop. And then pull it tight."

"What?"

"Here, I'll show you. It's not hard."

It's not long before the Mickey Mouse pole, and the Sleeping Beauty pole are all rigged and two volunteer, sacrificial worms are bobbing just below the surface of the water, nine or ten feet off the end of the dock.

After about 10 minutes of inactivity, Deb says, "This is fun," as she finishes a long, extended yawn.

"Nothing seems to be biting."

"Nope."

"I'm afraid my poor worm is drowning," I say. "Will a fish bite on a dead worm?"

"I don't know. Probably a more relevant question is whether or not a fish will bite on a Disney fishing pole."

"I think the answer is probably 'no' to both questions."

"You know, I feel sorry for these poor worms," says Mary. "Do you think they're in pain?"

"Hugh?"

"I mean, you push a hook through them three or four times and then you toss them in the water. How cruel. I wonder if they go to worm Heaven when they die."

"It's probably not Heaven. I think they all end up in worm Purgatory," says Mark.

"Poor little fellas."

I hold my breath for a moment, wishing that Mark had not used the 'Heaven' and 'Purgatory' words. We have all been getting along so well, and I dread the thought of reigniting another smoldering topic about religion. No sense in opening that can of worms. So, I think for a moment, searching my brain for another topic to talk about. But before I am able to formulate my thoughts, Mary takes the bait – a bait that I think Mark had intentionally dangled in front of her – just waiting for her to swallow the hook, so that he could reel her in.

"You know Mark," says Mary, "I have been thinking about all your Purgatory and Hades ideas. And I think you are wrong."

"Here we go again," I think to myself.

"I still believe that we all still go directly to Heaven when we die," continues Mary. "Or in your case Mark ... and I say this lovingly ... you may, possibly, go directly to Hell."

"I'm glad you said 'possibly' and not 'definitely'."

"Well, I'm not your judge, but I do think that your eternal fate is set at the moment you die. And I don't know any place where the Bible says that you go to some vague spiritual place where you work on becoming a better person. I don't remember seeing that anywhere. But the Bible does talk about going directly to Heaven ... and to Hell."

"And where does it say that?"

"Well, it's all in the story about Lazarus."

"Lazarus?"

"Yeah, Lazarus."

"You mean the guy Jesus raised from the dead?"

"Oh no. The other Lazarus. The story in the Bible where Lazarus goes to Heaven and some rich guy goes to Hell – right after they both bite the bullet. It's in the Bible."

And with that statement, I can't help myself. And I jump right into the conversation that I did not want to start in the first place, "Mary, you have got to stop taking the Bible so literally. It's just a book written 2000 years ago by a bunch of old guys trying to work out some type of belief structure, and I can't see how any rational person can believe that everything they wrote down is inspired by God."

I'm on a roll now, and I've got a platform and an audience, and so, even though it's against my better judgment, I continue. "You know, there are so many contradictions in the Bible – that you run into a lot of trouble when you try to believe everything – because you have to reconcile all the apparent discrepancies. And even if you believe that this good book is divinely inspired, not everything in it can be true. Two contradictory statements cannot be valid at the same time. I mean, how can you reconcile the Bible teaching that you should love your neighbor – and then have God tell Moses and the Israelites to murder all the pagan women and children when they rampage the former inhabitants of the Promised Land?"

"All I know," interrupts Mary, "is that the Bible says that all scripture is inspired by God and is profitable for doctrine."

"That's true", adds Mark, "the Bible does say that. But when Paul wrote that to Timothy he was really only talking about the Old Testament. Those were the 'scriptures' that were familiar to the Jews of that day. In fact, when Paul penned those words, a lot of the books in the New Testament were not even written – including the book of Revelation."

"Oh crap!" shouts Deb. "That sucker stole my bait. Matt, I need another worm."

"Would you like me to put it on the hook for you Honey?"

"Of course, Darling. Make sure you hook him good. And then be sure to wash all that slime off your hands."

"You know," says John, "that's something that has always bothered me.

"The slime?"

"No, the contradictions. I mean, there's a lot of talk about love and acceptance of others, but then the Bible comes down really hard on homosexuals."

"And adulterers too," I add. "Right Mark?"

"Oh yeah. I'm right there with John. The Bible is really hard on both of us."

"That's because you're both living in sin. And the Bible is always going to be tough on sin," replies Mary

"But it's not just the sinners," says John. "The Bible doesn't have a lot of nice things to say about women. In fact, women are not even allowed to talk in church meetings. They have to hold their tongues until they get home where they can pose any questions they have to their husbands – the husbands they are supposed to submit to."

"You have to realize," says Mark, "that the Bible was written in a different time – a time with different cultural norms and different social requirements. And much of the advice that Paul and other New Testament authors give to the early Christians – is not really applicable today. We live in a different culture. So I just ignore parts that are related to the social norms of the first century. The homosexuals, and women, and pagan children of today are so far removed from ancient customs and cultures – that we cannot force them to conform to the standards of the early Christians in Israel."

"So you don't believe anything in the Bible?"

"Oh no, I believe that the Bible has sound doctrine. It's a book that God wants us to have. To study. To learn. To live better lives. To begin to understand the nature of God and His creation. And I believe that there is something magical and mysterious about the Bible, the way that it all fits together, and the way it changes lives."

"But?"

"But, I just don't believe that it's infallible. The Bible says that scripture is inspired by God. It doesn't say that it's perfect."

"But you can't do that," says Mary. "You can't just strip out the portions of the Bible that you don't like and accept the rest. That doesn't work. If you do that, all you get is a hodgepodge religion of your own making. Something that God did not intend."

"That could be. And I struggle with that sometimes. But I think it is possible to build your faith on the total theology the Bible presents – without getting bogged down in all the social and cultural rules and traditions of the first century church."

"So being an adulterer is OK?"

"Well, I don't know."

"I think it's OK to be gay," says John. "It's the way I was born. God made me this way. I can't help it. I am His creation, and this is the way I turned out. So, given that, Mary, do you believe that all homosexuals are going to Hell?"

"Yes! Yes! Yes! Woooeeee! I got him! He's a big one!" shouts Deb as the tip of her pole bends down almost into the water.

We all watch with eager anticipation as Deb reels in her fish – all of us contributing to the celebration with a few hoots and whistles – all hoping to be the first to catch a glimpse of fish on the end of Deb's line.

And after about a 15 or 20 seconds of screaming and reeling, Deb pulls up ... a five inch long sunfish. Not exactly the prize catch we had imagined. But good enough for a few pictures taken with our phones, and a few good memories of a grand fishing expedition.

"We could always use that one for bait," says Matt.

“Go ahead and laugh Honey. We still have about 10 minutes before lunch, and that big one is still out there. So, I'm going to keep on fishing.”

“OK Darling.”

“You know,” says Mary. “I'm just fishing too. Just like Deb. I know I don't have all the answers. And I know I have a lot of studying to do before I understand a lot of what the Bible is teaching. So I'm just searching. And I do have an open mind and I am open to opinions that are different than mine. But I do believe in a more conservative interpretation of the Bible than you guys.”

“Mary, none of us has a monopoly on the truth. We are all fishing for answers – just like you. And we are all trying to figure out exactly what the Bible is teaching and how those teachings relate to the way we live our lives. No one has all the answers, and you know, there's always room for a difference of opinion.”

“Yeah, and so ... I keep coming back to Lazarus. You know, that story in the Bible where poor Lazarus sits at the gate of a rich man's mansion.”

“Yeah,” says Mark, “it's a parable that Jesus told.”

“Yeah, and Lazarus is starving and has sores all over his body, but the rich man refuses to help him.”

“And?”

“Well, they both die, and Lazarus goes to Heaven and the rich man goes to Hell. That's it. Case closed. That's what happens to you when you die.”

“Wait a minute. Not so fast,” says Mark. If you read that parable a little more closely, you find that it doesn't really say that Lazarus goes to Heaven.”

“Sure it does.”

“No. It actually says he goes to the bosom of Abraham.”

“Abraham has a bosom?” I ask.

"Well it probably just means he goes somewhere around the family of Abraham, but the Bible doesn't specify exactly where that is."

"So it could be in Hades or maybe Purgatory? I mean, didn't you say that that's where all the dead people go awaiting judgment?"

"Yeah," says Mark. "All the Old Testament characters are still sitting around waiting for Judgment Day – just like everyone else. But there is one more interesting point about this story. You see, the parable actually does tell us where the rich man ends up. And he doesn't go to Hell. The Bible specifically says he goes to Hades – possibly just down the street from Lazarus and Abraham's bosom."

"That just can't be," says Mary. "The parable says that Lazarus was with the angles – and the rich man was in torment. So they can't both be in Hades."

"I'm not so sure. You see, Lazarus and the rich man can see each other – and even talk to each other. So you would think that they might be in the same place. Now the Bible does say that they are separated by a great distance, and that there is a big ditch between them that they can't cross over, but I figure that Hades is just a pretty big place, and there are some cities in Hades that are better to live in than others – some that are fun and some that are a bit dreadful."

"Like Cancun being a lot better than Des Moines?"

"Maybe."

"I am going to have to check my Bible to be sure everything you are saying is true. I'm sure there are different ways to interpret what it says."

"Sure there are," says Mark. "And you have to realize that the story of Lazarus is just a parable, and you can only read so much into parables – because they are just stories – illustrations to explain a specific theological point. But if you still believe that Lazarus goes to Heaven, think of how painful it would be for him to look across a big canyon and be able to see and talk to family members and other loved ones that didn't make it – the ones that ended up in Hell. That would be a very painful experience, and not one that we would associate with being in Heaven."

"Yeah, that's a good point."

"There's the lunch bell," says Matt. "Time to eat. John, Deb, pull in your lines. We're done fishing for today."

We immediately begin picking up our empty water bottles and folding up our deck chairs and prepare to march off to the dining hall. It would be such a shame to miss a meal. But as John is reeling in his pole, and is just about ready to pull his line out of the water at the end of the dock, the float plunges beneath the water, and John's drag begins to scream as a fish pulls the line out toward the center of the lake.

"WoooEeeee," shouts Deb. "I don't think this is a sunfish. You've hooked a monster."

John's Mickey Mouse pole is almost bent in two, as he begins to tighten the drag, trying to turn the fish around before all the line on the reel is gone. But it is to no avail. The line snaps. And the fish is gone.

"That was fun," says John. "Too bad he got away. I wonder what kind of fish it was – and how much he weighed."

"Well, it could have been a large bass or possibly one of those big land-locked salmon. But it's hard to say. That's why fishing is so much fun – you don't always know what goes on beneath the surface. You never know what is lurking down there. I guess it's just like our discussion of Heaven and Hades and Purgatory – all places you can't see. When you have questions and go looking for answers, there will always be some fish that you will never be able to reel in."

After lunch we all begin our now familiar puzzle/shower round robin – where we alternately take showers and work on Deb's puzzle. And before long, we were all, once again, wearing towels, gathered around the puzzle, singing along with John's makeshift boom-box. Currently, an old Jim Croce song called "Bad, Bad, Leroy Brown" is playing.

And he's bad, bad, Leroy Brown
Baddest man in the whole damn town
Badder than old King Kong
Meaner than a junkyard dog

You might think that with all the practice we have had this week that our singing would have improved a smidgen. But alas, it appears to have had little effect. I guess practice doesn't make perfect.

"Here comes the best verse of all," says John.

Well the two men took to fightin'
And when they pulled them from the floor
Leroy looked like a jigsaw puzzle
With a couple of pieces gone

"That's appropriate," says Mary. "I think there are a couple of pieces missing from this puzzle. I'm not sure we're ever going to finish."

"Oh, I don't know, says Deb. "It's starting to come together. That cabin in the middle is almost complete, and it looks like there are some people standing in front of it – in the middle of all those trees and bushes."

"Yeah, it's getting there. I just wish there weren't so many green pieces that look alike."

"And blue pieces that look alike."

"And brown pieces that look alike."

"Yeah."

"Oh no," Matt says in a panic. "It's almost three o'clock. We've got to get ready. We can't go to the wedding in bath towels."

"We could," I respond, "but there might be ..." But before I am able to finish my sentence, everyone had already scampered off to their rooms to get dressed.

After rummaging through my suitcase, searching for some appropriate wedding attire, I put on a slightly wrinkled sport shirt, a pair of nice khaki shorts and a pair of white tennis shoes – formal enough, I think, for an outdoor wedding on an island in the middle of a lake. After all, it looks like it might rain again this afternoon, and this outfit will dry fairly quickly after a brief downpour.

But as I walk down the stairs, I notice that Mark and Mary have opted for more elegant apparel. Mary has on a colorful flowing dress, and Mark is wearing a sport jacket – all a bit overdressed for my taste, but I think that maybe, I should consider going back upstairs to change. And I am about to turn around when I overhear the conversation Mark and Mary are having in front of the fireplace.

"So Mary," says Mark, "are you still against this wedding? You know, the union of two heathens in a Christian church?"

"Oh, I don't know. I guess it's not really the wedding I object to. I mean, I think it's great that Ruth and Abdul are getting married. I just don't think that a church is the appropriate place for a Jew and a Muslim to unite forever in holy matrimony."

"I doubt that it will be forever."

"Well, maybe not. But you know what I mean. The church is all about Jesus, and forgiveness of sins, and salvation – and none of that applies here."

"So you don't think that there is any salvation for Jews and Muslims?"

"I'm afraid not. The Bible says that Jesus is the way, the truth, and the life. And that no one is able to get to God except through Jesus."

"Wait a minute Mary," says Mark. "It's really not that simple."

"Sure it is. The Bible says that God so loved the word that he gave his only begotten Son, that whosoever believes in Him should not perish but have everlasting life. It's right there in John 3:16."

I really don't want to get involved in another one of these theological conversations, especially with Mary quoting that 3:16 verse again, so for the moment, I decide to keep my mouth shut and let Mark somehow convince Mary that all non-Christians are not going to Hell. I know he is much more knowledgeable on this subject than me, and I know that if I somehow relent and enter this conversation, and spout out my usual "when-you-die-it's-over" philosophy – everyone will just get angry. So, to avoid any confrontation, I simply walk over to the bar and pour myself a straight scotch and take a couple of big gulps.

Before this week I was convinced that there really wasn't anything to this "life after death" thing. And I believed that religions that teach that you pass through to Nirvana, or Heaven, or some Happy Hunting Ground after you die – were just ... well, mistaken. To me, the concept of one's soul surviving a bodily death just didn't make logical or scientific sense. I mean, how can any intelligent person living in the twenty-first century believe in ghosts?

But over the last few days, I have begun to rethink many of my long held beliefs. I keep hearing Matt say that things are not always as they appear, and my mind has become a jumbled mess with thoughts of black holes, and time, and consciousness. And now, I'm not sure what I really believe. Maybe I'm just too emotionally involved right now. Maybe the stress of this family reunion and the news of my returning cancer has affected my normal, rational thought process. Maybe after I get back home to my everyday routines everything will be clearer.

But for now, I'll just try to cope the best I can. And I guzzle a little more scotch as Mark continues his conversation with Mary.

"So Mary, what about Abraham?" asks Mark.

"The Old Testament Abraham?"

"Yeah. That one. The father of Judaism. The one that lived way before Jesus. Is there any chance of salvation for him? I mean, he was never born again – never able to profess his faith in Jesus – and never saved before he died. So, is he excluded from this eternal life thing – just because he was born a few thousand years too early?"

"It's different for people in Old Testament times."

"How's that?"

"I don't know. It's just different."

"And what about babies that die in infancy?"

"I don't know how it all works, but I know that God is kind and just, and so I'm sure he'll find a path to salvation for innocent babies ... and for everyone that loved God before Jesus was born."

"Well, if God is kind and just, what about the poor little orphan boy on the streets of Calcutta who has never had the opportunity to hear about Jesus? Is he excluded from the everlasting life club?"

"I don't know for sure. But that's why we need to go to India, to Calcutta, and tell everyone there about the good news of Jesus."

"Yeah," says Mark, "and Christians have done that all over the world. But what if we just speculate for a moment, and say that there might be some path to salvation for people who have never heard of Jesus – just like babies who die before they are able to comprehend the Bible's good news."

"OK. But I don't think the Bible talks about that."

"No it doesn't, but just follow along with me for a moment."

"OK."

"So, say we find a tribe of South American Indians living in the Amazon jungle who have never had any contact with the outside world – never even heard of the name 'Jesus'. And so let's imagine that maybe, just maybe, there's a way for our 'kind and just' God to save their souls."

"Yeah. I'm sure God will be kind and just to anyone who has never heard about Jesus."

I think I know where Mark is leading with this conversation, and I chug the rest of my drink and pour another glass as Mark continues.

"Well, let's say we get together a group of missionaries, rent a couple of dugout canoes, brave the elements, the piranha, and the malaria infected mosquitoes, and go down and find these poor people in the middle of the jungle ... and tell them about Jesus. What then?"

"Well, then they are assured of their salvation."

"Yeah, but what if they reject the message? What if they don't turn from their old ways – from the beliefs and customs of their ancestors? What if they send you away refusing to convert to this strange white man's religion?"

"Well, then they are condemned," says Mary."

Mary may not be following Mark's train of thought, but I know what's coming next. And I know that Mary has been trapped into a theological corner. And I take a few more sips of scotch as I wait for the trap to close.

"OK," says Mark. "So they're condemned. But don't you see. It's your fault. The reason they're condemned is because you brought them the message of the gospel. If they had never heard that message, if you had never gone into that jungle – they might have had a chance at salvation. Just like that newborn baby that died who had never heard about God. They might have been saved because they had never had any exposure to the Bible and its teachings. But because you gave them a choice, because you told them about Jesus, and perhaps, because you just didn't make a convincing enough argument ... they have no chance at eternal life. And it's your fault. Because of you, they are worse off now than they were before."

I think about entering the conversation at this point – to say something profound about Heaven and Hell and eternal life, and how society has distorted our perceptions and understanding of those concepts, but I am not able to come up with a way to express my thoughts clearly on the subject. I'm a little confused, I guess, about what I actually do believe. And so, I just listen, and take another sip of my drink.

"That's a flawed argument," replies Mary after a brief pause. "If that were true, if telling someone about Jesus would condemn them to Hell, then it would be better if no one ever talked to anyone else about Jesus. And that's just stupid."

"I agree," says Mark. "The argument is flawed. I'm just not exactly sure where the flaw is."

"I think there's a flaw in this bow-tie," says John as he walks down the steps. "I can't seem to get it even on both sides."

"You're wearing a tie?" I ask.

"I'm not," says Deb as she and Matt walk out from their room. "I've got shorts on just like you Luke. But, I made Matt dress up a little."

"Just a clean shirt and some slacks," says Matt. "Come on. We better go. Don't want to be late for the wedding."

I feel a little tipsy as we walk down the steps to the dock, and I stumble a little stepping into the boat. A little too much scotch, I guess.

"Luke," says Matt, "you better let me drive. You can sit in the front and watch out for rocks."

"OK," I reply, actually glad to have someone else taking the helm.

The sky is now completely overcast and the wind has picked up considerably from this morning – now blowing steadily from the northeast with intermittent strong gusts that rustle the trees and stir the water.

"Peter said that a front may come through this afternoon," says John. "I just hope the rain holds off until after the wedding."

"Me too."

We cast off from the dock and head off toward Church Island, but with all five of us in the boat, all sporting a little additional weight from all the meals we have overindulged in this week, we have a little trouble getting our small dingy up onto a plane. I walk up to the front of the boat to try to hold down the bow as the stern sinks deeper into the water as we begin to accelerate, but the tilt of the boat causes me to lose my balance for a moment and I almost fall overboard. Fortunately, Deb, just in time, grabs the back of my shorts and pulls me back down into my seat.

"Better let me sit in the front," says Deb.

We arrive at Church Island without further incident, and disembark from the boat with the help of a couple friendly greeters stationed on the dock. "Just follow the path to the chapel," one of them says. "We should be starting in just a few minutes."

In double file, we walk down the dirt and pine needle path toward the chapel. The path is lined with plaques that have pictures of Camp Chocorua, the former boy's camp which was established on this island in the late 1800's. There's a picture from 1882 showing the sailing kayaks built by the boys to navigate the lake. And a little further down, a photo from 1885 showing the Chocorua Island Boys Choir dressed in their black robes and white smocks, walking double file, down the path to the chapel, the same path we are on right now – a path that has changed very little in over a hundred years. The inscription on the photo reads, "One hears in the distance clear young voices ... and along a path through the woods, with the sunlight dancing in and out among the branches, the boys come nearer."

"It's almost as if you can see and feel those boys, walking with us down the path right now," says Deb.

"Yeah, sure," I reply.

A little further along, we pass a picture of the first wedding performed in the chapel in 1916, and then, a picture of a church service conducted in 1949 – with all the women wearing appropriately coordinated long dresses and wide brim bonnets.

As we reach the rear of the chapel, we stop for a moment to admire the scene, with the majestic white birch cross standing prominently in the front of the chapel, and the serene lake in the background. "This is beautiful," says John. "it's amazing how this little relic of the past has survived for so long – through two world wars, and then through Korea and Viet Nam and Desert Storm."

"Yeah," says Deb. "It even made it through the 70's – and disco."

Most of the wedding guests have apparently already arrived, and we all take a seat in the back on the left side of the chapel – the bride's side, which is made up of a mostly Jewish brigade, with many of the men sporting yarmulkes on their balding heads and the women wearing stylish "Neiman Marcus" type dresses.

Across the center isle from us is the Muslim battalion – where the mostly dark skinned middle-eastern contingent has their base camp. Some of the women are wearing a hijab, the traditional Muslim veil that covers the head and chest of women, which is typically worn whenever an adult Muslim woman is in the presence of adult males.

There is no interaction between these two companies on the two opposite sides of the chapel – with both detachments talking mostly among themselves – and with the center isle playing the role of a de-facto demilitarized zone – separating the West Bank from the Hebrew territories. And you can almost feel the tension and distrust emanating from the troops on both sides of that line of demarcation. "What a shame," I think to myself. "I just hope Nan can pull off this wedding with no incidents – no rockets fired into enemy territory, or cross-border kidnappings, or improvised explosive devices ruining the festivities. It would break her heart if her only daughter's wedding turned into a sectarian confrontation between wedding guests."

After a few tense moments, the organist breaks the apparent impasse and begins the process of detente by walking down the center isle to the small building to the left of the altar. He then opens the doors to the organ, pulls out the bench, sits down and begins arranging his music. In the meantime, two young boys run around to the side of the building and begin turning the organ crank – pumping a refreshing blast of fresh air through the organ pipes. The organist begins by playing some classical piece, Bach I think, which seems to ameliorate some of the tension. It's funny how music can have such a calming effect. And we all begin to feel a little more at ease as the ushers begin to seat the families of the bride and groom.

Nan had asked me earlier during the wedding planning process if I would like to sit with her in the front of the chapel with the bridal party, but I graciously declined. Sitting up there would mean that I would have to be in the same isle with Nan's ex-husband, and that was something that would have been intolerable for me. We have never gotten along, I think mostly because he is jealous of Nan and my relationship. It has always been my impression that he believed that someday he and Nan would get back together – even though Nan has made it clear to him that that will never happen. But he has never given up trying, and every time we meet, he goes out of his way to needle me and to belittle my abilities and accomplishments. He is vicious and condescending, and is a master at verbal abuse. For many years, I tolerated his attacks, until one day, after I had had more than enough, and I decided never again to have any interaction with him. So, I just cut him off – out of my life completely, and I have not talked to him since. I don't even refer to him by his name anymore. I just call him "That Pompous Fat-Ass". Nan says that I should be a little more understanding, but there are times when she gets as frustrated with him as I do.

Nan is one of the first to come down the aisle, arm in arm with a young usher. I want to get a picture as she walks by and I pull a small camera I brought along out of my pocket and step over Mary and John to the center isle in an attempt to capture a memory on disk. But in my haste, I trip over a rock next to John's chair and fall on the ground, causing the poor Jewish lady in front of us to scream as I almost knock her out of her chair.

Nan just gives me a disgusted look as the commotion settles down, and then continues walking down the aisle with her good-looking escort, as I sheepishly return to my chair. "Sorry mam."

The rest of the wedding goes along pretty much the same as any other wedding – the bride coming down the aisle with "That Pompous Fat-Ass" as the organ is playing

Pachelbel's Canon – followed by a young girl singing some nice song – and then the usual ceremonial stuff.

I start to dose off for a moment but wake up as the Justice of the Peace gives a nice little chat acknowledging the 800 pound gorilla in the room – saying that the union taking place today is a union of two great faiths, and that this marriage is an eloquent expression of the hope that all mankind has for the future – an example of how the world can live together in harmony.

I think that this little dissertation on hope for the future is a bit sappy and trite, but as I look around, everyone else seems to be receptive to his message. I guess, it was a good thing to say given the audience. Anyway, as I sit there, I start hoping this guy would just finish with his comments so that we can get this wedding over with.

But then, right before the vows are taken, a stunning sequence of random events occur. It starts when a bald eagle which has been circling overhead, for some reason, decides to attend this marriage ceremony and lands in a tree just to the right of the altar. And he just sits there, and reverently watches the proceedings taking place below, with a stately demeanor and presence that these beautiful creatures always seem to have. It's as if he came to sanctify the ceremony – to offer his encouragement, and to give his approval to the union taking place.

And immediately a song comes to my mind. It's called "On Eagles Wings". It's a song we used to sing in church when I was a kid. And I begin humming it quietly underneath my breath.

And He will raise you up on eagles' wings
Bear you on the breath of dawn
Make you to shine like the sun
And hold you in the palm of His hand.

And just about that time, directly above the chapel cross, two large dark rain clouds part, and a ray of sunshine illuminates the area around the altar.

This whole sequence of events could not have been choreographed any better – the eagle landing in the tree and watching over the ceremony, and then the altar being bathed in sunlight just as the vows are recited. It's as if God himself is giving His approval of this union. It seems as if He is sending a message to everyone that all will be well, and that this ceremony is sanctioned, and is blessed by the Almighty himself.

"I now pronounce you man and wife. You may kiss the bride."

And with that, a thunderous applause breaks out throughout the chapel – on both sides of the aisle. And it seems as if the barriers that once divided and separated these two culturally diverse groups has somehow been broken down, as the two sides meet and greet each other in the center aisle, with smiles and handshakes and back pats.

"I guess there is hope," I think to myself. "Maybe this marriage can work. Maybe this is a lesson, an example for all of us. Maybe we can all learn to be more tolerant and accepting of others. It's really not that hard to get along with people that are different than you. It's just like our family, who throughout this week has learned to put aside our differences and nurture the common bonds that hold us together."

But as the celebration begins to wind down, and everyone begins to file out the rear of the chapel ... the clouds above close, and the sun is once again hidden behind a dark grey veil. And the eagle, perhaps frightened by the commotion – flies away. And after a few minutes, a light rain begins to fall, and everyone scampers back to their boats, to seek cover before the real storm strikes.

The Reception

A light but steady rain continues to fall as we motor back to our Deephaven dock, and by the time we arrive at the cabin we are all wet and chilled. We think about starting a fire in the fireplace but decide that there is not enough time, because we have to leave for the reception. So, we hastily change clothes and pile into Mark's rental car and head off to the party.

The reception is at Nan's hotel which is about a fifteen minute drive around the lake from our cabin. When we arrive, the festivities have already begun. A four piece band is playing, the champagne is flowing and the hors d'oeuvres are being ransacked.

We all queue up in the reception line to greet the bride and groom, and some of the family members. Everyone looks a little soggy from the rain, but otherwise in good spirits. I exchange a few pleasantries with Ruth and Abul and then move down the line to say hello to Nan.

"You look beautiful," I say as I give Nan a small kiss on the cheek.

"Thanks," she replies. And then she looks to her left to greet the next person in line as I pass on to the other family members.

I had expected a warmer welcome from Nan after not seeing each other for a few days, "But, I guess that's alright," I think to myself. "I'm sure she's stressed. I'll be able to talk to her later. She'll be a little more relaxed after a couple glasses of champagne."

At the end of the reception line is Nan's ex-husband, and I think for a moment that I will just pass by him without saying anything, but then I have second thoughts, and in a spirit of acceptance and reconciliation, I say, "Saul," using his actual name instead of calling him by his 'Pompous-Fat-Ass' nickname, "it was a beautiful wedding."

"Yeah it was," he replies. "Except for when you screwed up the opening by falling down in middle aisle in front of the whole procession."

"I'm sorry. I tripped."

"I bet you're sorry. You ought to be – making a spectacle of yourself like that."

"It was an accident, and it really didn't affect the wedding very much. And I am truly sorry that I caused a commotion."

"Yeah, I can't believe you would do something like that to Nan on such an important occasion. How embarrassing for her. But I'm glad to hear that you at least have some remorse – that you are 'SORRY'."

"I am sorry."

"Yeah, I'm sure you are sorry," says Saul as he chuckles under his breath. "In fact, as far as I can tell, your whole life has just been one sorry excuse after another. I don't know what Nan sees in you. From everything I have seen and heard, you're really just one big loser."

And that causes me to lose my temper.

In my anger, the muscles in my right hand spontaneously contract forming a tight fist, and my arm simultaneously rears back, into a position of power and leverage, preparing to flatten this son-of-a-bitch with a knockout punch.

But just as I am about to lunge forward, and teach this obstinate a-hole a lesson about making inappropriate remarks to his wedding guests, I realize that I am about to ruin the entire wedding reception – a reception that means so much to Nan. And so, I stop myself and halt the attack. But as I pull back and stop the fist that is headed for this guy's jaw, the momentum causes my whole body to move forward, and I stop right in front of Saul's face, our noses just a few inches apart. And being unable to remain in such an uncomfortable position for long, just staring into those hardened beady eyes, I instinctively put both hands on Saul's shoulders, and push him away. Not a hard push. Just a small, light shove, but a shove that I will regret for a long time to come.

Saul falls back two steps to maintain his balance, but in the process, left his foot hits the leg of a chair behind him, and he stumbles, and turns and falls. And as my luck would have it, he falls on top of a table set up directly behind the reception line – and by some twisted quirk of fate, this particular table has a cake on it. A four layer cake with Corinthian style columns holding up each layer. A cake with intricately swirled white frosting precisely patterned in crisscrossed designs with occasional leaves and flower buds. A cake capped with two figurines of a well-appointed bride and groom, happily strolling through the icing on the top layer.

It's the wedding cake that Nan had spent many hours designing. It was so important to get it just right, because it's the cake that Nan had wanted for her own wedding before she settled for a more modest version because of a hastily arranged ceremony and a limited budget. It's a cake that means so much more than just a few pounds of flour and sugar, because it was Nan's gift to her daughter. A gift that was intended to convey a message – the message that Nan approved of this union between Ruth and the man she has chosen to live with the rest of her life, a message that she believes that love can overcome racial and religious conflicts, and the message that this marriage has Nan's blessing and support.

And I watch helplessly as the symbol that was meant to sanction this marriage is lost for all time. Saul falls face first directly on top of the cake, and the table collapses, and Saul, and the broken figurines, and the cake – Nan's loving gift to her daughter – all lie battered on the floor.

Nan runs over to help Saul up. And I, not sure what to do or to say, just manage to stutter a few words, "Nan, I, I, I, I'm so sor-sor-sorry."

And Nan turns her head toward me, and with tears running down her cheeks, says, "Just get out of here. Now!"

"Can I have another drink?" I ask Mark who is attending bar back at our cabin.

"Sure."

"It's OK Luke. Don't beat yourself up too much over this," says Matt as he throws another log on the fire and sits down in a rocking chair next to me. "It was an accident. You had no way of knowing that that big doofuss was going to trip over his own fat feet."

"But it was my fault," I say, "I pushed him."

"Yeah," says John, "but I heard what he said to you. He deliberately provoked a confrontation. I would have slugged him in the nose."

"It probably would have been better if I had hit him. It might have spared the cake."

"Maybe," says Mark, "but then I may not have gotten all these great pictures. Look, here's one of Saul licking the icing off his face. It's classic."

"That's not funny."

"Sorry."

"You know," I say as I slump back in my chair, "I wanted so badly for this wedding to be perfect. For Nan's sake. It meant so much to her. And I was worried that one of the wedding guests from either the Jewish or the Muslim contingent was going to spoil the celebration. But it was me. I ruined the whole thing."

"Oh I don't think you ruined the whole celebration. The party was still going strong when we made our hasty retreat out the back door. The music was still playing and the champagne was flowing. And they still had the groom's cake to cut. I think the reception was just fine. The cake incident was just a minor distraction."

"Maybe. But I fear that Nan will never forgive me. Lately our relationship has not been the best. She has been increasingly critical of me and the dumb things that I do sometimes. Honestly, I have always wondered why she has put up with me all these years."

"I know she loves you very much."

"I'm really not sure about that any more. And I'm afraid that tonight I might have pushed her over the edge. You heard what she said when we were leaving."

"She didn't mean that. She was just upset."

"I'm not so sure. I just hope that, somehow, Nan will give me a second chance. Just an opportunity to make amends."

"Take it from me," says Mark, "with women, you almost always get a second chance. It's the third and the forth chance that become tricky and a little more challenging."

"Great."

"You know," says Deb, "I feel as if my whole life has been a second chance – and opportunity to amend all the bad things I did in my former life."

"What?"

"Yeah. I don't think I was a very good person. In my last life, that is. And now, I've been given the opportunity to do better. A second chance."

"You had a former life?" asks John, a bit surprised by Deb's statement.

"Deb," interrupts Matt, "it might better if you didn't talk about this. I'm not sure everyone will really understand."

"Oh, it's OK. They're family."

"So Deb," says Mark pursuing the odd conversation we had just entered into, "Who were you? The Wicked Witch of the East or the Wicked Witch of the West?"

"Oh, I really don't know. It's all kinda vague. I just get glimpses every once in a while of my former self. And somehow I know that the reason I'm here now, in this life, is to work on becoming a better person – taking the lessons I have learned from the past and growing into a more loving and caring individual."

"That's ridiculous," I say. You only have one life. What you think are glimpses into your past, are only figments of an active imagination. It's not hard for anyone to conger up images of ourselves in the past – playing out roles that we have seen in old movies or read about in history books. It happens to everyone. I have a recurring dream where I'm Larry Bird making a last second buzzer beating shot to win an NBA championship."

"We all know that could never happen," injects Mark.

"Yeah. But you see what I mean. It's just our mind playing games. Don't you agree Matt?" And when Matt doesn't respond for a few seconds, I repeat, "Matt?"

"Well," says Matt finally, "there have been a number of studies done where people under deep hypnosis recount stories of former lives. And many of the stories are very compelling, because many of the former life experiences that these people remember help to explain many of the physical and psychological problems they have in their current lives."

"What does that mean?"

"Well, patients that died because of falling off high cliffs will often have an overwhelming fear of heights in this life, and people that had some type of severe wound will sometimes have birthmarks or birth defects in the same place as those former injuries."

"But that's just ..."

"Yes, your right. I know what you are going to say. You really can't rely too much on what someone says under hypnosis. They are very vulnerable to suggestion and the mind can create some very real and compelling imaginary images."

"So you think most of the stories these people relate under hypnosis are just dream induced fantasies?"

"Oh, I'm not really sure. I often wonder why so much of our personalities are ingrained into us, into our DNA, before we are born, and I wonder how the kids that we call child prodigies can so easily accomplish some very difficult tasks. And so, I can't say for certain that these individuals did not have some type of prior life experience. And the reason that I can believe in that possibility is because I know that the perceptions we have of this world do not give us an accurate representation of actual reality."

"But the Bible says that you only have one life," says Mary.

"Yeah, that's true," replies Mark. "But you have to be careful. When the Bible talks about life and death, you are never really sure about whether it is talking about physical life and death or spiritual life and death. And besides, there's always John the Baptist."

"What does John the Baptist have to do with this?"

"Well," says Mark, "John the Baptist was reincarnated."

"What?"

"Yeah. It's true. And we have that on pretty good authority," says Mark. "Jesus said so."

"Really?"

"Yeah. And the Bible is pretty clear on that point, and it's not subject to a lot of misinterpretation. Jesus says that John the Baptist is Elijah, the Old Testament prophet whose spirit has come back in the body of John. And that statement is recorded three times in the Bible. Twice in the Gospel of Matthew and once in the Gospel of Mark."

"They never taught us that in Sunday School."

"No. Because it doesn't fit into the little box of beliefs that most people have about Christianity. To most of them, reincarnation is strictly a Hindu or Buddhist belief – not something that a good Presbyterian believes in. But that's what the Bible says."

"So Mark," says John, "how do you think this whole thing works? I mean after you die. You think your soul goes to this Hades or Purgatory place where all the dead people are, and you work on becoming a better person and then maybe come back to earth to try to live a better life?"

"Oh I don't know. That's a bit of a stretch. But like we talked about before, there's really not much in the Bible that says anything about what happens to us immediately after we die. There's a long period of time between when we kick the bucket and when we face our final judgment – when we get the thumbs up to enter the pearly gates. And what happens during that interim period has been a matter of speculation for centuries."

"So, do you think that we may get a second chance?" asks Mary.

"What do you mean?"

"You know. Like we talked about this morning. All the babies – and the isolated Amazon Indians – and the little kid on the streets of Calcutta – and all the Old Testament guys – the people that have never heard about Jesus. Do you think they have a second chance? An opportunity sometime after they die to accept Jesus as their Savior?"

"Perhaps. I don't know of any place in the Bible that says that we don't get a mulligan."

"So all the Hell and brimstone preachers are all full of hot air?"

"Maybe. It's hard to say. But you know, I'm counting on having a second chance. I know that my life has not been exemplary, and I look forward to the opportunity to improve on my shortcomings."

"Why don't you start working on them now, Mark?" asks Mary.

“Well,” replies Mark, “I kinda like my life the way it is right now. And it's hard to change a lifestyle you enjoy. So, I just postpone abandoning all of my vices until a later time.”

“So it's OK to screw all the women in your former congregation now because you feel like you'll get an opportunity to repent of your bad behavior sometime later?”

“That's harsh.”

“Well,” says Mary, “that's the fallacy of your argument here. If we all get a second chance to follow Jesus after we die, then there is no reason to modify our sinful behaviors in this lifetime. And how would all that work anyway – with everyone doing what they please, because it really doesn't matter – because we will all have an opportunity to live a better life later. That just makes no sense at all.”

I am about to enter this conversation – to contribute my two cents worth, but fortunately, we are interrupted by a loud noise coming from the corner of the room. We all look over – and see Deb – slumped over in her chair – snoring louder than any proper woman should be allowed.

“I guess I need to get Deb to bed – before drool starts running down her face,” says Matt.

“Good idea. I think it's time we all hit the hay.”

“Yup. Good night everyone.”

“Good night.”

“Good night.”

CHAPTER 6 – DAY 5

Reality is merely an illusion, albeit a very persistent one.
Albert Einstein

Unfinished Business

I had a little trouble walking back from breakfast – just could not seem to catch my breath. I had to stop at the ice house on the way for a brief rest, and when I finally do make it back to the cabin I collapse in a chair next to the fireplace. I'm not sure why I am feeling this way. Perhaps it is because I didn't sleep very well last night – thinking about Nan and worrying about what I am going to say to her today when she returns from the hotel. Or, maybe the weakness I am experiencing is somehow due to my cancer. Perhaps it's more advanced than I think. Or maybe it's just in my head – all psychological. "Yeah," I think to myself. "That's probably it."

"Why don't you come over here and help me with the puzzle Luke?" says Deb. "We're going to try to finish this thing today."

"Maybe in a little while."

"This puzzle is starting to come together," says Mary as she scans the table looking for a piece with a particular shade of green. "I can't wait to finish it. We think the cabin in the center may be one of the cabins somewhere around here on the lake and we thought we would kayak around a little this afternoon to try to figure out which one it is."

"I may just stay here and take a nap," I reply, and I slouch down in the chair and rest my head on one of the pillows. But just as I find a comfortable position, sitting sideways with my knees pulled up, almost in a fetal position, the front screen door squeaks open and Peter walks in.

"Good morning everyone! Beautiful day. Isn't it?"

"Hello Peter. How are you doing?"

"Oh just fine. I brought you some fresh ice for your icebox ... and Luke, I have a couple of phone messages that were left for you at the main office."

"Oh?"

"Yeah, the first is from your wife. She said that she has a couple of loose ends to tie up and won't be back to Deephaven until sometime tomorrow."

"Oh."

"And the second one is from Stephen's family."

"How is Stephen?" asks Matt.

"Well, it's really a miracle. Yesterday they weren't expecting him to live through the night – but early this morning he snapped out of his coma and sat up in bed."

"That's great."

"Yeah, he's still in critical condition. Lots of internal organ damage. Especially his heart, but he apparently is in good spirits. He said that it wasn't quite his time to go yet ... because he had some unfinished business."

"I hope he doesn't want to go back into the water and finish his swim in the lake," says Deb.

"No," replies Stephen, "but he asked to talk to Luke. And he was wondering if you could come to the hospital. Today, if possible."

"He must be confused," I say. "I wasn't the one that saved him. It was John and Matt. I wasn't even there. I think you guys should go. Not me."

"No," replies Peter. "I talked to Stephen's wife on the phone, and she was very specific. He wants to see you Luke."

"OK. But I don't understand why he would want to talk to me. I'm not even sure how he knows my name. We were never formally introduced. The only time we ever talked to each other was when we were trading cabins, and so, I'm not sure visiting him in the hospital will accomplish much. But I can go, I guess – if that's what he wants."

"It's a bit of a drive. The hospital is in Lebanon, about an hour and a half from here."

"Oh, that's OK. It doesn't look like Nan is coming back today. So, I don't have much else to do."

"You know," says Matt, "maybe you should wait until tomorrow to go. It's a beautiful day and we were planning to hike up Rattlesnake Mountain this afternoon. It's so clear today. The view from up top should be spectacular. We may be able to see all the way to Lake Winnipesaukee."

"I'm not sure I'm up for a hike today," I reply. "I'm really tired."

"Well," says Deb, "if you don't go on the hike ... and you don't go to the hospital ... you can always stay here and work on the puzzle."

"Yeah. I guess I can. So Peter, can you give me directions? I think I'll leave right after lunch."

"Sure."

"You can use our car," says Mark.

"Thanks."

Solitude

One of the reasons I had agreed to drive over to see Stephen at the hospital was because I needed some time on my own. By myself.

For some reason, solitude is comforting for me. I can retreat into myself, and not have to be concerned about what others are doing or thinking – or about whether I am acting appropriately in a social situation. Life is easier when the only person you have to worry about is yourself – when you don't have to think about impressing the boss, or carpooling the kids, or helping that old lady cross the street. There's a certain peace that comes from knowing that the only person you are responsible to – the only person you have to answer to ... is yourself.

I realize, of course, that my longing for solitude is one of my biggest problems – that when I shun social situations I miss out on any opportunity to build the relationships that are so meaningful to most people – relationships that, I know, could be meaningful to me as well, if I were just, somehow, able to get closer to others.

But with intimacy comes vulnerability. And vulnerability ultimately leads to pain.

The most fulfilling and memorable times of my life were when I was involved in a strong, close relationship. We were so happy – my wife, me, and our darling three year old daughter.

That is, until that one day, when the bond that we had forged together was broken. And just like a nuclear fission reaction, the pain unleashed from the severing of that bond was overwhelming. A mushroom cloud of sorrow and suffering, destroying everything in its path. And in the end, my wife was gone, my career was in shambles, and my daughter was dead.

And I often wonder, if I would have been better off never to have had a daughter. Never had those glorious moments when she said "Da Da" or when she took her first steps, or when she would run into my arms when I came home from medical school. It was only because of those events, that relationship, that bond ... that the pain of losing her was so great.

It's a sad fact of life. The more you love, the more you ultimately suffer.

As I leave Deephaven, driving down the narrow, winding, tree lined road that circles the lake, I think I may stop to see Nan. I'll be driving through Holderness on my way to the hospital, and I could drop by for a few minutes – to see if Nan needs any help, and to apologize for my behavior at the reception, and to tell her that I have missed her, and that I love her.

And so, when I reach the stoplight in Holderness, the only stoplight in town, I put on my left turn signal – to turn towards the hotel where Nan is staying. But when the light turns green, I turn right instead, and head out of town.

Right now, I just need some time by myself.

Over the Rainbow

It's a beautiful drive to the hospital – through the rolling hills on small two lane winding roads that often have no shoulders. I pass through several pristine forested areas, which then give way to open pastures with small lakes and quaint old farm houses next to old barns with rusted doors and silos that seem to be tilting slightly to the side. There are herds of cows waiting patiently for their turn in the milk shed, and endless fields of bailed hay – so freshly cut that their aroma permeates the air inside the car.

I slow down to the posted speed limit through the small towns of Danbury and Grafton, enjoying the view as I pass by. But I become a little too distracted and miss a turn in Canaan and go quite a few miles out of my way before I realize my mistake and turn around. But that's OK. I'm enjoying the drive, and I roll down all the car windows and let the wind blow through my hair as the sights and sounds and smells of the country pass by.

I arrive at the hospital after about two hours, and the friendly volunteers at the front desk give me Stephen's room number and point me toward the elevators.

Stephen is in a room with two patient beds, and so I quietly begin walking past the first bed to get to Stephen on the far side of the room. But before I get too far, a head pops out from beneath the covers of the first bed, and a small boy says, "Mister, can you help me?"

I keep walking for a moment, not really wanting to get involved with this little tot, but then he says again, "Hey you. Mister. Yeah you. Can you help me?"

And I turn, and I see this small kid waving a remote control back and forth in the air. "This thing won't work," he says. "It won't play my movie. You need to fix it."

It seemed like a simple request that I should be able to take care of expeditiously, so I walk over to his side and ask, “What seems to be the problem?”

“I can't get the dumb movie to start, and they've got all these stupid tubes in my arm and I can't get out of bed to fix it.”

I smile at the little boy's assertiveness, and walk over to the side of the room and open the tray in the DVD player and straighten and reinsert the movie disk. And in a few seconds the movie begins to play on the small screen above the kid's bed.

“Oh, *The Wizard of Oz*. One of my favorite movies,” I say.

“Yeah. Mine too. It's a little scary. But I'm a big boy. I can handle it.”

“I'm sure you can.”

“Thanks mister.”

I then walk around the curtain and see Stephen. He is apparently asleep, but if it weren't for the raspy breathing noises he is making I would have thought he was dead. He is so pale and unresponsive, with a ghostly stark white face and deep sunken eyes. He is lying on his back with both hands folded over his heart – looking more like a corpse than a recovering patient. And my stomach becomes a little queasy when I see all the tubes and wires and needles and monitors that surround Stephen's bed. He looks so much worse than I expected, and I try not to disturb him as I take a seat in the chair beside his bed, and listen to the monitors make strange beeping noises as wiggly lines scroll across the various LED screens.

While I am sitting there I can hear the movie playing on the other side of the curtain.

Dorothy? Well, what has Dorothy done?

What has she done? I'm all but lame from the bite on my leg.

Oh! You mean she bit you?

No, her dog.

Oh, she bit her dog, eh?

And, after a few minutes, I am just about ready to doze off in my chair when I hear a faint, weak voice say, "Luke. Thank you for coming. I'm glad you're here."`21

"Hi Stephen. We have all been worried about you. How are you feeling?"

"Oh, I guess I'm OK. I just feel kinda stupid for standing out in the middle of the lake right before a big thunderstorm."

Now look here, Dorothy, you ain't using your head. You'd think you didn't have any brains at all. Your head ain't made of straw you know.

"Yeah, I guess we all should have known better. Is that the last thing you remember? I mean, standing on the swim platform before the lighting hit."

"No. It's not," says Stephen as he coughs a few times, which causes his monitors to momentarily beep louder and irregularly. "And that's what I wanted to talk to you about. Do you think you can hand me some water please?"

"Sure."

Stephen takes a few gulps from the hospital sippy cup and continues, "I know you are going to think that what I am about to tell you is hard to believe. But you have to hear me out. So please, just humor me, and follow along."

"OK," I respond cautiously.

"You know, I remember standing there, on the platform, looking out at the horizon, thinking that we had a good ten or fifteen minutes before the storm was going to hit."

Better get under cover, Sylvester. There's a storm blowin' up – a real whopper.

"And then I remember the lightning flash, but I never heard the thunder."

"It all happened so quickly," I say. "I looked out of the cabin window right after the flash, and I saw you in the water."

"Yeah, but you see, after the lightning strike ... after my body fell into the water ... well ... I know it's hard to understand ... but that guy in the lake wasn't me. I wasn't in the water. My body was there. But it wasn't me."

"What?"

"It wasn't me, repeats Stephen. "And I know that because I was watching everything. I saw it all happening. From above. I watched everything happening from above."

Somewhere, over the rainbow, way up high.

I feel compelled to interrupt Stephen's story at this point – to explain that his visions were probably just the wild imaginings of a brain starved of oxygen, but I hold my tongue and don't say anything.

And the dreams that you dare to dream
Really do come true.

And Stephen continues, "And I saw Matt capsize when he tried to climb into his kayak. And I saw Mary and Deb fall into the water when the sailboat mask falls on top of their canoe. And I saw your dock collapse when Mark forgets to untie the rope on the motorboat."

I realize now, that Stephen's family must have told him about our rescue efforts, and somehow, in his mind, Stephen had become confused, and thought that he actually witnessed our failed rescue attempt. I guess that's understandable, given the trauma he's been through. So, I continue listening.

"And I keep wondering. Why is everyone so concerned about that body in the water? I'm up here. And I'm fine."

Stephen coughs a few more times, and then gasps for breath, a raspy breath that sounds like there is fluid building up in his lungs, and I reach over and adjust the oxygen tube in his nose which had become dislodged.

Stephen closes his eyes and takes a few short breaths of the fresh oxygen. Then he pushes his head back slightly against his pillow and, within a few seconds, falls asleep. I check his monitors to be sure he's still alive – but there doesn't seem to be any alarms beeping or unusual lights flashing. So, after a few minutes, I begin to get up out of my chair to leave the room. But then abruptly, Stephen wakes up – and continues telling his story – not missing a beat – as if he had never dozed off.

"And the next thing I knew, I was in a long tunnel, traveling at a very high speed – faster than I have ever gone before – faster than you can even imagine – traveling through that tunnel."

It's a twister! It's a twister!

"And then suddenly, I was surrounded by a brilliant light. Brighter and whiter than anything I have ever seen. And I feel peace and comfort and love in the light. Lots of love."

Toto, I've a feeling we're not in Kansas anymore.

"And then I am face to face with this being. A being of light. And I am encompassed by the light – as it surrounds me completely."

Are you a good witch or a bad witch?

This whole thing is beginning to feel a little awkward, and I squirm in my chair a little, and look over to the door of the room – trying to formulate an exit strategy – figuring out how to leave gracefully without insulting Stephen or upsetting him in his tenuous medical condition. But Stephen just keeps on talking.

"And I am consumed completely by the light, and I sense and feel this being's presence, and love and acceptance ... and forgiveness. You see, he knew everything I had done in the past – both good and bad. And he told me things. And he laid out for me the path that I am destined to follow in the future."

Follow the yellow brick road.
Follow, follow, follow, follow, follow the yellow brick road.

"And Luke, that's why I needed to talk to you."

By now, I have had about as much as I can take of Stephen's imaginings, and I feel that if I stay any longer this fairy tale is just going to become weirder than it already is. And even though I want to bring Stephen back down to earth – to knock some sense back into him – to bring him back to reality ... at this time, I don't have the heart to tell him that this is all just a dream – that his guy in the light is just an apparition – a figment of a damaged brain, trying to make sense of a very unexpected and traumatic event.

I do believe in spooks! I do, I do, I do, I do!

Ah! You'll believe in more than that before I'm finished with you.

So, it's time to go. I have done my part. I feel as if I have fulfilled my responsibility. I have come to see Stephen, as he requested, and now, I need to leave this place. This fantasy about the light isn't going to get any better, and I have the impression that if I continue listening, that Stephen will probably start talking about meeting Jesus or something. So, I get up out of my chair and slowly begin backing out the room.

"Well, Stephen," I say timidly, "that's a fascinating story ... but you must be exhausted. You need to get some rest. Thank you so much for sharing all that with me."

I can't really believe that I used that "sharing with me" cliché. I'm not sure I have ever said that before, but it seemed like an appropriate way to make a hasty exit – without seeming rude. And I turn and begin to walk out of the room.

And I am almost past the curtain dividing the room when Stephen coughs – a huge rasping cough. And then he says, in a horse and barely audible voice, "Oh Luke. Just one more thing."

I think for a moment that I will keep walking and pretend I did not hear Stephen's faint plea, but for some reason, I turn, and take a few steps back toward the bed.

"Yeah," I say.

And what Stephen tells me next ... leaves me speechless.

I get lost driving back to Deephaven, and I spend quite a while driving aimlessly down back country roads. I'm having a little trouble concentrating on where I am going, I guess. Actually, I'm having trouble concentrating on anything. My mind is just a jumble of meandering thoughts and distorted images – trying to piece together everything that has happened. And when I finally arrive at the cabin around sunset, exhausted, I remember very little about the drive home – the route I took, or the turns I made, or the cars I passed. I don't know how fast I was going or whether or not I was low on gas, or whether the radio was on or off. Everything is just a fog.

When I stumble back into the cabin, I am greeted by Matt.

“Oh, there you are,” he says. “Looks like you missed dinner Luke. But I brought you back a sandwich. Hope you like tuna salad.”

“Thanks.”

“How's Stephen doing?” asks John. “Is he OK?”

“Yeah, fine,” I reply – not wanting, at the moment, to begin a lengthy conversation about my visit to the hospital. And I slowly walk over and collapse into that comfortable rocking chair next to the fire.

“That's great. We were all worried about him.”

“Yeah.”

“Luke,” says Mark, “you missed a great hike today. We climbed to the top of Rattlesnake Mountain. It was beautiful up there. So clear. You could see forever.”

“Yeah,” says John. “It was my first time up there. You guys would never take me along with you when we were kids. It's something I have always wanted to do. And it wasn't really that hard a climb, except for that steep part toward the end.”

"I almost fell backward on the steep part," says Mary. "But it was worth it. What a view. Just standing there, looking out over God's wonderful creation. It's funny how you feel so much closer to God when you're up there. It's as if God's spirit is there admiring the beauty right along with you."

"Really?"

"Yeah, it was almost a spiritual experience," adds Deb. "Sorry you had to miss it Luke."

"Yeah," I reply almost in a whisper.

"Luke, are you OK? Is there something wrong? What's the matter?"

"Oh," I respond, "I just need to talk to you guys for a few minutes."

"What's the matter?" asks Deb as everyone sits down around me in front of the fireplace.

I had originally thought that I would keep all the details about my conversation with Stephen to myself – no need to involve the rest of my family in a matter that really pertains only to me. But for some reason, I feel the compulsion to open up a little – to expose myself to the vulnerability that comes along with talking about your weaknesses.

And so, I begin to explain, as best I can, what happened today at the hospital. But I start slowly and cautiously, not knowing exactly how far I will go – how much I am willing to reveal. And I start with a simple question.

"Have any of you ever read any of those books that talk about near death experiences?"

"I've read a couple of them," says Deb. "It's fascinating reading. And I believe all of them. I believe everything they say. You know, about having out of body experiences and seeing the light and all that."

"What about you Matt?" I ask my erudite sibling. "What do you think?"

"Well," replies Matt, "it's really hard to say. Lots of people think that the visions people have when they are near death are simply the imaginings of a hypoxic brain – a brain deprived of oxygen where the neurons are slowly dying, and the synaptic connections between all those nerve cells that channel the electrical connections in our heads – become dysfunctional and confused. And so, you end up with short circuits, along unexplored neural pathways. And those short circuits lead to some pretty strange and exotic dreams."

"So you don't believe these experiences are real?" asks Mark.

"Oh no. I didn't say that. I just said that that's what many people believe. But I have learned over the years that some pretty strange things can happen, and that our limited perception of the world around us can mislead us into thinking that there are rational explanations for everything we experience."

"So you think they are real?"

"I'm not sure. There was one guy that admitted that he made up all the experiences he recounted in his near-death memoir, but many of the other books seem to be genuine."

"I've read some of those books," says John, "and a lot of people believe they are true. They say that if the visions were due to a dying brain, that they would be confused and muddled and erratic. But over and over, the accounts these people describe are very lucid and very specific and very real."

"Yeah," says Deb, "and they're very universal."

"What do you mean?"

"Well, Americans and Europeans and Asians all have very similar experiences."

"Really?"

"Yeah, and so do children. They all go through, pretty much, the same series of events."

"What events?" I ask.

"Well, almost everyone describes having some type of out of body experience, where they're floating, watching their lifeless bodies or other events occurring around them. And for many people with these experiences, it ends there. And they are reunited with their bodies and that's all."

"But what about the rest? What happens to them?" I ask.

"Well," says Deb. "Many people travel to another place, usually at very high speeds either in or through a long tunnel."

"What do you mean another place?"

"Well, it's usually a place of brilliant lights. That's why all the books that are written by people with these experiences always have the word 'light' in the title. They are either bathed, or surrounded, or saved, or renewed by this light. It's like I said. They all have the same type of stories. And this light that cloaks them is peaceful and loving and joyful – and almost everyone that progresses to this stage describes a beautiful visual and sensory experience in this light."

"And is that where it ends?" asks Mary.

"Sometimes, but some people talk about meeting deceased relatives, and some meet light beings – people made entirely out of light. And some have life reviews where they see all the good and bad things they have ever done in their life."

"Like Santa's list."

"Kinda. And some people are given instructions – things they need to know or things they need to accomplish after they return to their bodies."

"But they all do return to their bodies?" asks Mary. "Don't they?"

"Yeah, if they didn't I guess it wouldn't be a 'near' death experience. It would just be a death experience."

"Good point."

"So Luke," says John, "why did you ask the question? Why did you want to know about these near death experiences? Is that what Stephen wanted to talk to you about? Did he have one of these things?"

I just sit there for a few moments – not knowing exactly what to say or how to proceed. I'm having a little trouble processing everything Deb has just said – about the different stages that these people experience – and how similar the experiences are for everyone, even for people in faraway places.

"Luke, did Stephen have one of these experiences?"

And after another pause, I say, "Yeah. He did." And then, very slowly and methodically, in almost a trance-like state, I say, "And he was floating above the lake ... and he witnessed our failed rescue attempt ... and he went through that tunnel ... and he came to that place of brilliant lights ... and he saw that being of light. All exactly like you described."

"Wow."

"You know," says Mark, "I'm not sure near death experiences are all that uncommon. Paul had one."

"Paul who?"

"You know, the apostle Paul."

"You mean in the Bible?"

"Yeah, that's the one. And he writes about it in Second Corinthians. Somewhere around the twelfth chapter, I think."

"Really?"

"Yeah, Paul writes about his experience in the third person. He says that he knows a man who was taken up into the third heaven – to paradise. But most Bible scholars believe he was taking about himself because of the context of the verses. And Paul says that he heard things there that cannot be told. Things that no man can utter."

"Third heaven?"

"Yeah. Most of the ancients believed that there were three layers to the heavens. The first heaven was where the birds were, and the second heaven was the realm of the sun and the stars – and the third heaven was where God lived. So saying that he went to the third heaven was just Paul's way of saying that he made it to the top – as high as you can go."

"So was Paul hit by lightning too?"

"No, but he was close to death on a number of occasions. He was whipped five times, beaten by rods on three different occasions and shipwrecked three times. But, it is most likely that he had his near death experience when he was stoned and left for dead in Lystra – in present day Turkey, during his first missionary journey."

"They never taught us that in Sunday school."

"No."

"So, getting back to Stephen, you said he saw us trying to rescue him?" asks John.

"Yeah ... and he saw Matt flip over in his kayak ... and Mary and Deb fall out of the canoe ... and Mark collapse the dock ... and Matt and John pulling his body to shore on the water bike ... and ..."

"And?"

I have to swallow a couple of time to clear my throat, and then I continue. "And then he saw me."

"But you weren't there," says John

"No, I was. I was in my room. Watching out the window. And Stephen saw me ... and saw the debilitating fear I was experiencing ... saw that I was scared for his life ... and that I was scared for my own. You see, somehow ... he knew about my cancer. He knew that my lymphoma had returned ... even though I had just found out myself. "

"Really?"

"Yeah ... and he told me that I don't have to be afraid of dying ... that dying isn't the end ... that life goes on ... ," and then my voice begins to crack and I have trouble continuing.

"What a wonderful message," says Mary. "Maybe that's why Stephen had to come back. To tell you that. To give you hope. It must be why he asked to see you."

"Maybe."

"You know the tunnel that Stephen traveled through?" says Matt. "Well I find that aspect of these experiences very interesting – where you're traveling at very high speeds through a long tunnel."

"Why is that?"

"Because that experience is very similar to what you would expect if you were falling into a black hole. You see, the gravitational forces in a black hole are so strong that it severely bends all the space around it. And it bends that space toward the center of the hole – so much so, that it forms something very similar to a tunnel. I know it's hard to visualize, but imagine you have a slingshot with a really stretchy rubber band. If you pull that rubber band back three or four yards – then there is a very long narrow area between the body of the slingshot and the end of that stretched out rubber band. And that's kinda like how space is stretched in toward the center of a black hole. It forms a long narrow tunnel."

"Are you saying that you can get to Heaven – or Hades or Purgatory or where ever ... by falling into a black hole?" asks Mary.

"Oh no. Absolutely not. I'm just saying that it's an interesting parallel. Especially since you're traveling at very high speeds and end up at a place with lots of light. I never really thought about it before, but it would be fun to look at all the similarities."

"Yeah. Well," I say, "I'm tired. It's been a long day. I think I'll turn in for the night. Thanks guys. Thanks for listening to my story. Thanks for being here."

"Wait a minute!" says Deb. "You can't go to bed yet. You have to tell us the rest. You have to tell us what else Stephen said."

"No, that was everything. The floating thing, the tunnel, the light guys. That was about it. There wasn't much else."

"Luke. Come on. I know there's more."

"How do you know that?"

"I just know. What else did he say?"

The lump in my throat comes back, and I can feel my eyes beginning to water. And I look around the room, and everyone is staring at me. Just waiting. Hanging onto the next words I am about to say. And so I try to retreat.

"It was nothing important. Nothing that would mean anything to any of you. Just a few personal things."

"Luke, I know it's hard for you. But this is one time that you need to open up. You need to allow us to become involved in your life. We're family. And we can't act like a family – we can't function as a family – unless you are willing to let us in – to let us help you. And now is that time," says Deb in an unusually firm and animated tone. "Talk to us."

"OK," I say as a single tear begins to roll down my left cheek. "You're right. Stephen did say one other thing." And then I pause for a moment, trying to collect my thoughts. And then I continue, "Do any of you remember my daughter's name?"

"Oh, it was so long ago," says Matt. Wasn't it Lizzy or Libby or something like that?"

"No, but that's close. It was Elizabeth. But we always called her Liz. But I doubt that anyone knows her other nickname. The name we always called her because of her favorite food. Does anyone know?"

I wait for a few seconds, and no one answers, and so I continue. "We called her 'Peanut' because she was so small, and because she loved peanut butter so much." And then I swallow hard, and I continue as I begin to cry. "And Stephen knew her name. And he said that Peanut loves me very much. And that she misses me."

And that causes everyone in the room to gasp, as they try to comprehend what I have just said. No one says anything at first – not sure how to respond until the silence is broken by Mary, who begins to cry along with me, and says through her sobbing, "Luke, that's wonderful."

"And then," I continue, "he assured me that Peanut is doing just fine where she is right now, but that she is looking forward to see me sometime soon."

And then we all huddle around the fire together. All of us. Arm in arm. And we hold on to each other, tightly, no one saying anything, but no one letting go. And I feel a warmth and a love and a companionship between all of us. It's a feeling I have not had in a very long time, and I feel compelled to tell the rest of the story. And so, I break the silence.

"And then Stephen said one more thing. He said that it wasn't my fault."

"Wasn't your fault? What wasn't your fault?"

"Don't you see? I was right there. I saw the car coming. The car that hit her ... and I could have saved her ... but I froze ... I was paralyzed ... I couldn't move ... I wasn't able to react ... just like I couldn't react in the tunnel when the brakes failed ... or when I saw Stephen sinking beneath the water ... and in those few seconds where I was not able to move ... I watched helplessly as the car smashed into her tricycle."

"But Luke," says Mark, "It happened so fast. You can't blame yourself."

"Yeah."

"And Luke," says Deb, "don't you see? All these years you have carried this burden with you – never telling anyone. You held it all inside the whole time – never releasing that guilt. No wonder it's been so hard for you. No wonder the pain has never gone away."

"Yeah."

"But you have to understand Luke," continues Deb. "Stephen was right. It wasn't your fault. There wasn't enough time for anyone to save your daughter."

"Yeah. That's what Stephen said. He said that if I had run out into the street – that if I had made an attempt to save her – that we both would have been killed. And Stephen said that that wasn't the plan."

"What does that mean?" asks John.

"I'm not sure. That's just what he said."

And we all continue to stand there, arm in arm, for a long time – until the fire slowly burns out, and then we say good night, and go off to our rooms, to bed.

I have been trying for hours to fall asleep, but it's no use. I just roll back and forth in bed, from my right side, to my back, to my left side, and then back again.

I just can't stop thinking about everything Stephen said. And then, as I roll back to the right, I wonder, again, how he could possibly have known about Peanut. And how it could be that my little daughter is waiting for me? How could that be?

Then I roll to the left and think that maybe, it's just a trick – a cruel hoax pulled off by Stephen and a host of co-conspirators – spending months maliciously searching for little known names and facts about me and my daughter. You can find almost anything on the internet these days.

And rolling to my right, I realize that his family must be in on the whole thing. I'm convinced of that. They must have told Stephen about the rescue attempt – and about how we pulled Stephen out of the water. And they must have, somehow, seen me in my room – I know it's hard to see anything through the window screens – but they may have had some special-vision glasses, or there may have been some type of small video monitoring device hidden in here – maybe in the closet, peering out through a knot hole in the wall. That's got to be it. They must have heard my conversation with the doctor. That's how they found out about the cancer.

Or, I think as I roll onto my back, maybe there really is something to this life-after-death thing. Could it be that all those religious fanatics been right all along? No, it just can't be. I refuse to believe that those odd-ball, Bible toting, street corner preachers are anything but crazy lunatics.

But maybe, I think as I prop up a second pillow behind my head, maybe there's a reasonable, scientific explanation for all of this. Maybe your consciousness isn't confined to the axons in your brain. Maybe it can go other places – somehow travel through this space-time-continuum thing – where time and space stretches and shrinks, and where time can stand still. I know it sounds odd, but maybe there is a place where your consciousness, or your spirit, or whatever – goes after you die. But how could that be?

And I roll onto my right side and pound the pillow twice with my fist.

CHAPTER 7 – DAY 6

Science without religion is lame, religion without science is blind.

Albert Einstein

An Early Morning Climb

I open my eyes, not sure if I had ever really fallen asleep, and stare, once again, at the luminescent dial on my watch. It's 4:07am. Three minutes since the last time I looked.

So I close my eyes once more, trying to clear my mind – blank out everything, and think about nothing, so that I can fall asleep. But then, I think about what Mary said last night around the fire. She was talking about Rattlesnake Mountain, and how beautiful the view was on top. And I remember that she said that up there, she felt closer to God – that God's spirit was up there, standing there, on top of the mountain with her.

And it is at that moment, that I realize what I have to do. I have to climb that mountain. That's where the answers I am looking for must be – up there with God's spirit. And I have to get there before dawn, because there's something about light that I need to understand, and somehow, I'm not sure why, I know that I need to experience the sun's first rays as they beam from the horizon to the top of the summit. The same rays of light that Stephen spoke of. The brilliant rays that surrounded and consumed him – the light that emanated so much love and acceptance and forgiveness.

It's that light that I need to see. Firsthand. The light at the top of the mountain – where God's spirit roams. The answers must be there – somewhere in the light.

There's no question in my mind. I need to go. I must climb the mountain, and I must climb it now.

I hastily but quietly get dressed and walk down the stairs and out the front door, being careful to slowly close the screen door on the porch, so that the squeaking spring will not wake anyone. I bring along a small flashlight with me, but the moon is almost full and is so bright that I put the flashlight back in my pocket.

"I guess I won't need this tonight," I whisper to my moon shadow.

It's about a fifteen minute walk to the base of the mountain, down Pinehurst Road to the beginning of the trail, where I stop to rest for a few minutes.

"I'm already out of breath," I think to myself. "I just don't have the stamina I used to. Maybe this is a bad idea. I don't know if I can make it to the top."

And I sit down on a large granite rock, and prop my head up with my hands while resting my elbows on my knees – and I begin to face the realization that my cancer has returned, and that it is taking over – ravishing my body – progressing farther and faster than ever before – and sapping all my strength and energy as it invades my liver, and my bones and perhaps my brain.

I don't know how much time I have left. But I know it's worse than I had imagined. I may be near the end. My life may be over. "What a waste," I think to myself. I could have accomplished so much, but now it appears that my final resume will just be cluttered with failures, disappointments and lost opportunities.

And it's much too late to change things. I don't have the time or the strength to change course – to make something better of my life.

I should probably go back to the cabin and rest. I need to forget about climbing the mountain. The light I am searching for is beyond my reach.

But then I take a deep breath, and forge ahead. I have to try.

"Just one step at a time," I think to myself. "The right foot followed by the left – again and again – up the mountain."

I grab a discarded walking stick from the side of the trail to help me up the path, and I climb higher and higher by anchoring my feet on exposed tree roots crossing the washed out trail, and on granite stones that were randomly left behind from the last ice age. My cramped legs are on fire, and my breathing is strained and erratic, but I am gaining ground.

"Just one step at a time."

I have to stop every five or ten minutes for a short rest, but then I continue forward.

"I can do this," I repeat over and over to myself, and my confidence is building as I finally near the top of the trail. But then, once again, like so many times before, fate strikes another blow.

Suddenly, my left foot slips off a moss covered rock and my ankle twists in the soft clay on the side of the trail. And I fall back, grabbing a small sapling to try to regain my balance. For a split second, I think I have averted a disastrous fall, as I cling precariously to the small tree, holding my breath as a single bead of sweat drips off my brow.

But the mountain is conspiring against me. She knew I was coming, and had carefully placed the slippery rock and the soft clay along the way, and now, she has given me a young tree much too weak to support my weight. The sapling breaks, and I careen backwards, head first into an immovable century old oak. The majestic tree holds me up for a moment, but then gives me back to the mountain. I collapse to the ground, semi-conscious, and roll down the slope, finally being stopped by an old rotted log crossing the path.

And as I lie there on the damp, dew soaked ground, I realize that my time has run out. I don't have the strength to go any farther. I have fallen for the last time. And I rest my head on the soft, leaf covered ground and close my eyes.

I am ready to die.

"Com'on Luke. Let's get up. It's not your time yet."

I open my eyes. I am dazed for a moment, not able to remember where I am. But as my blurred vision begins to sharpen, and the aching in my head and ankle become more acute, I remember my failed climb, and I realize I am lying across the trail, blocking the path for everyone else.

I see in front of me a short portly gentleman, with a white beard, wearing overalls and a lumberjack shirt. It seems lighter now, as I can clearly see his weather worn face and his benevolent expression. It is the type of face that is hard to forget – strong rustic features, but with pudgy cheeks and kind eyes.

"How long have I been here?" I wondered.

"Looks like you had a pretty bad fall Luke, but I think you're gonna be OK."

"Who is this guy," I think to myself, "and how does he know my name?"

"It's not that much farther to the top. I think you can make it. Let's go."

The kindly gentleman helps me up, and as I reach my feet, to my amazement, I feel better. Maybe the fall was not that bad after all. Strength has returned to my legs and my lungs feel refreshed in the cool morning air. I manage to speak for the first time, "You from the camp?"

"Oh no," said the gentleman, "but I've lived in these here parts for a long long time. Ready?"

He puts my arm on his and we begin walking the same path that had defeated me earlier. Seems easier now though. It's amazing what a little rest can do to rejuvenate the body. We walk past the rock and the clay and the sapling, defying the mountain's previous advances against me – up the steepest part of the trail with minimal trouble. And as we continue our climb, my new friend looks over at me and smiles, with those soft eyes and rosy cheeks, and he nods, as if to say that he is enjoying my company. And I begin to feel a sense of joy, and peace and accomplishment.

And then, after sidestepping some loose rocks, we round the final bend on the trail, and I can see the clearing at the top of the mountain. And as we pass the last trees, I turn with eager anticipation to view the majestic lake below.

"Seems darker now," I think to myself. "I can't really see much." I then look to the horizon and notice just a faint glimmer of light in the east. The sun is just beginning to rise. "I made it! I made it!" I mumble to myself out loud. Finally, for the first time, I'll be able to see the sun rise from this precipice – to witness the majesty of God's creation as soft rays of light slowly illuminate His handiwork. And perhaps, God's spirit will show up as well – somewhere in the light – right alongside of me. And so I wait – motionless, as if in a trance, as I stare out toward the horizon – and watch, completely mesmerized, as hews of red, orange and yellow begin to emerge in the eastern sky.

As the light becomes stronger, I begin to see the clear calm waters of the lake, reflecting the sky above and the trees below. So peaceful. So surreal.

"If there is a God, he must be here somewhere. If only I knew where."

I subconsciously find myself repeating a verse I learned as a child in Sunday school.

He leadeth me beside the still waters.

It's from some Psalm. I don't remember exactly which one, but I know they always recite it at funerals – and I have been thinking a lot about funerals lately.

The rising sun warms my body, as I soak in its energy and strength.

He restoreth my soul.

A light breeze has begun to blow across the lake causing tiny waves on the water. The sun's rays landing on these ripples create an array of millions of small star-like reflections across the pond. The dazzling display makes we wonder, "Why is the light so comforting? Is it just its radiance, or perhaps the warmth it brings – or is there more? They say God is light. I know that's just a metaphor, but maybe there's more to it."

This week has given me a new appreciation for the power of light, or maybe the power in the light. I close my eyes and turn my face toward the sun, trying to feel the energy – the life in the light. But after a few minutes, I turn away, unable to sense anything out of the ordinary. I have learned so much, but there is still so much that I don't understand.

There is, however, one thing I do know. I no longer have an overwhelming fear of death. Dying doesn't scare me as much as it once did.

Yea, though I walk through the valley of the shadow of death, I will fear no evil.

I begin thinking about my funeral. "What are people going to say? And why does it matter anyway. I won't be there. I'll be in ... well, somewhere. Maybe, I guess. I'm not really sure anymore."

... and I will dwell in the house of the Lord forever.

There are so many things going through my mind. Up until this week I had always thought that once you died, it was over. Just a decaying body to be buried or burned. And that's it. Dust to dust. No immortality. No eternal soul. Done.

Now, however, I'm confused. I'm a practical guy, and my training in the natural sciences and my medical background had taught me to believe that science could explain all things, given time. But now I don't know what I believe.

I close my eyes for a moment, and try to construct a prayer asking God, or someone, or anyone – for understanding, or help, or comfort. But I become frustrated with the attempt. I can't find the right words. "This is useless," I think to myself, "I don't know what to ask for. I don't even know who I am trying to talk to."

As I open my eyes, my attention returns to the lake, and as I peer down, I notice how different things appear when you look at them from on high. There is a completely different view of the world from above.

I keep hearing Matt say, “Things aren’t always as they appear,” over and over again.

I am beginning to realize that everything looks different when you change your perspective. This week I have come to appreciate that the things we see, the things we experience, the things that we are sure of – may not be an accurate description of reality. I am going to have to start looking at things from a different vantage point – from higher up on the mountain.

And so, I sit down on a rock, and peer out on God's creation – a creation molded and formed over millions of years – mountains rising through volcanic and seismic activity, and valleys carved by giant sheets of ice, and animals evolving from simple aquatic creatures to fish to amphibians to reptiles to dinosaurs to eagles. And I wonder, “What role does God play in all this? Is this place's beauty only due to natural selection and random mutation, or is there more to it?”

And as I continue to stare out toward the horizon, mesmerized by the beauty unfolding in front of me as the sun rises higher in the sky, bathing me in its ever strengthening light, I think that maybe, just maybe, science and religion can learn to get along – to coexist – even though they so often seem at odds with each other.

“Scientific Religion,” I think to myself. “Now that's an oxymoron if I've ever heard one.”

But then I start thinking about what Deb said. We were talking about consciousness, and discussing how this self-awareness might just be a phenomena of the electrical activity of an evolutionary advanced brain, when Deb said something that I will never forget – something that I keep hearing her repeat over and over again in my mind.

She said that the molecules in your head don't fall in love.

And I begin wondering about our consciousness. What is it? How does it work? And why is it so difficult for scientists to study? I remember Matt talking about the binding problem – how somehow our brain is able form images of what we see and experience by amalgamating inputs from all over our heads – dredging up memories and feelings from the past and assimilating them with sensory inputs from our eyes and ears and nose which have spread out to various parts of our brains – and finally, forming an image inside our cranium – an image that we see and experience, but an image that is formed where it is completely dark. I never really thought about it like that before.

And then Matt said something very simple and profound. He said that one of the reasons consciousness is hard to understand is because no one has ever been able to look at a 'feeling' under a microscope. He always has such a marvelous way of explaining difficult concepts.

Except for maybe Quantum Mechanics. That still has me confused. But I guess I'm in good company – even Einstein thought Quantum Mechanics was spooky.

But even so, I have to wonder, how is it that our conscious observation of photons or electrons or cats or whatever – can cause events to change? How can it be that the random firing of neurons inside our head can affect the world around us? I mean, how can it be that just looking at Schrödinger's cat creates its reality – either a live or a dead kitty? It just doesn't make sense.

But then, nothing Deb does or says makes much sense either. What is this connection that she has? How can she sense things like she does? How did she know that we weren't in any danger in the tunnel when the breaks on the van failed? And how did she know that trouble was brewing before the lightning struck Stephen?

And so, I wonder again, what is consciousness? Is it just some type of neural activity between axons in our head, or can our consciousness really communicate with the world around us? Deb said that consciousness was our spirit, and that this spirit gives us our memories, feelings and emotions. And so, I wonder, if this spirit is not limited by the axons and synapses in our scull, well then maybe, there exists the possibility that our consciousness, our self-awareness – can exist outside our bodies – and that it can continue on after our death – after all those billions of brain cells die.

I can't believe that I am actually thinking about these things.

But I keep hearing Matt say over and over again, "Things are not always as they appear."

And I keep wondering whether or not someone can believe in both science and religion. Can a rational 21st century adult believe in the Big Bang and evolution and black holes – and at the same time believe in the archaic teachings in the Bible? Two things that have always seemed so incompatible to me.

But maybe my understanding of the Bible is flawed. Maybe I have too many misconceptions of what the Bible actually says. And apparently, many other people have misconceptions as well – misconceptions about Heaven and Hell, and about the rapture, and about judgment and salvation.

I have always viewed the Bible as just a bunch of rules and regulations. And if you followed the rules you went to Heaven – if not, to Hell. But that always seemed so arbitrary to me. How could a just and loving God condemn so many of His children to eternal torment – especially all those who were born into social situations where they had little or no opportunity to hear or understand Christian theology? Millions of people – people created by God – all destined to burn in the eternal fires of Hell because of the social-economic conditions into which they were born. All the Muslims on earth, and the Buddhists and Hindus, and all the Jews. All doomed.

And what about babies who die in infancy?

And all the people that lived before Jesus was even born?

How could that be ... unless ... unless you get a second chance.

Mark says that the Bible says very little about what happens to you right after you die, and that judgment doesn't take place until sometime way off into the future.

And so, I begin to wonder, if your consciousness does continue on after our death, and we end up in Hades, or Purgatory or that paradise place that Mark has been talking about, what happens then? I don't know. But maybe you have a chance to make something better of your life. Lord knows I could use a second chance. My life has pretty much been one disaster after another. I have made more than my share of bad choices and poor decisions. But maybe I can have a fresh start – a new beginning – a second chance.

Mark said that in the Bible, Elijah came back to earth as John the Baptist. Perhaps we can all do the same thing. But maybe not. From what I know of Elijah, he was a pretty good guy – a man of God – a prophet. He probably had some type of elite status that allows him special privileges – and I'm not even a member of a church.

Matt said that people under hypnosis sometimes tell stories about former lives, but I'm not sure you can believe all that. I went to a show once where a guy hypnotized people in the audience, and he was able to get them to do and say anything he suggested.

So, I'm confused. There are so many different thoughts going through my mind right now, and I really don't know what I believe anymore.

Maybe I need to look at all this more scientifically. What can science tell us? At one time I thought I understood that question, but now, things are not as clear as they once were.

I'm still confused about this space-time continuum thing. I mean, how can space and time stretch and shrink? And how can time pass slower for someone traveling really fast – or someone in a strong gravitational field? And how is it that time stops completely in a black hole? I'm not even sure I understand exactly what that means. Time stopping. It seems like a contradiction of terms – another oxymoron. But if time does stop ... is that what eternity is? Or heaven? Or everlasting life? Everything going on forever and ever? It's almost too abstract to comprehend.

And then there's space. Empty space. A vacuum that somehow bends and stretches – even though it's just nothing – complete emptiness. But somehow, there might be six or seven extra dimensions curled up inside this emptiness. I can understand our three dimensions, height, length and width, but I cannot even begin to imagine how to add a fourth, or a fifth. How does that work?

But if there are other dimensions, what's inside them? Parallel universes? Wormholes to other places? Heaven and Hell – or perhaps Hades? I don't know. I guess no one does.

But let's get back to the facts. The things that I understand. The things I have experienced and that I know are true. I know that Stephen was unconscious. I saw that firsthand. He was hit by lightning, he fell into the water and he drowned – and he didn't regain consciousness until a couple of days later at the hospital.

But somehow he knew.

He knew the details about our rescue attempt. He knew I was in my room in the cabin. He knew that I was distraught after hearing that my cancer had returned. He knew about my paralysis.

... And he knew about Peanut. How could that be?

No one else – no one in the whole world – knew about the guilt I carried for so many years. No one knew that I panicked – that I froze when I saw that tricycle roll into the street. No one knew that I have always felt as if I could have saved her – if I had just reacted sooner. I kept that to myself – all these years.

But Stephen knew. And that's a fact.

And now, I need to reconcile that fact with my concept of how the world works – with my personal belief system, which, up to now, has not allowed any room for God or for eternity or for life after death.

As difficult as it is for me to believe, somehow, there must be something to this near-death experience thing. So much has been written about it. There must be some inkling of truth there – somewhere. I'm not sure I have any other explanation. All the stories these people tell are so similar. It can't just be a coincidence that so many different individuals from so many different backgrounds and cultures – have the same experiences – where they find themselves floating above their bodies, and traveling through a tunnel, and encountering a light – the exact same experiences that Stephen had.

The same type of experience that the Apostle Paul had – over two thousand years ago.

So, there must be something to it. But how could that be? I wish that I had the answers – but I don't. I need help, because right now, I don't know what to believe – and I have nowhere to turn.

And so I pray – the first prayer I have uttered in a long, long time – a simple prayer – short and to the point, "Please, God, help me to understand."

And almost instantly, I hear a voice from the distance, "Luke! Luke!"

It's a familiar voice. One I recognize instantly. And I turn, and I see her. It's Nan, running toward me with outstretched arms. A glorious sight. And we embrace – a long, strong embrace.

We don't say anything to each other for a moment, just enjoying the prolonged hug, until Nan squeezes me a little harder and whispers, "We were all so worried about you Luke. Are you alright?"

"Yeah, I'm fine."

"They told me about Peanut."

I don't reply for a moment – and Nan continues, "And they told me about your cancer."

"They did?"

"Yeah," says Nan as she looks directly into my eyes, "but everything will be OK. We have each other. I love you so much."

And then we both begin to cry. And as we are standing there, bathed in the warm morning sunlight, we are joined by Matt, Deb, Mark, Mary and John who wrap their arms around the two of us and extend our embrace. And we all stand there. Arm in arm. In the light. Our family. Together.

"It looks like we're down to our last few bottles of wine," says Mark as he begins pouring everyone a glass.

"Who drank all that stuff?" asks John.

"I don't know. It just kinda evaporated. Good thing this is our last evening."

"I really don't want this week to end," says Mary. "We've had such a good time. We'll have to plan to get together again. Soon."

"Yeah," I reply as I take a sip of a vintage glass of Bordeaux. "I'll be in touch with everyone, and try I'll arrange a reunion that fits into all our schedules."

"Great!"

"Do we have any appetizers to go along with the wine?"

"Just some crackers and some bread. Looks like that's all we have left."

"That will do just fine," says Deb as she looks up from the puzzle. "It's only about a half hour before dinner."

"How's the puzzle progressing Deb?"

"Oh, well, I'm afraid I have some bad news for everyone."

"What?"

"Well," replies Deb as she brushes a few cracker crumbs off the table, "It doesn't look like we will ever be able to finish the puzzle."

"Why is that?" I ask with my mouth still full of bread.

"Because, it looks like we are missing a bunch of pieces. That's the reason it has been so difficult finish. A lot of the pieces that we have been looking for just aren't here."

"Did you check the closet where you found the puzzle?"

"Yeah. Nothing there. And I looked under the rugs, and on the bookshelves and in the microwave too."

"We have a microwave?"

"No, John. That was a joke."

"Oh."

"That's a shame," says Mark. "Now we'll never know what the finished puzzle looks like."

"That's not really true," says Deb as she looks up from the puzzle and takes a gulp of her wine. "There's enough here to go on, even though we can't see the complete picture. We know that the cabin in the center is probably one of the cabins here at Deephaven. And we know there are six, seven or maybe eight people in the front. We can't tell exactly what they are doing, but they seem to be enjoying themselves."

"Isn't that what was hand written on the paper bag you found the puzzle in?" asks Nan. "Didn't it say something like 'Our Family'?"

"Yeah, it said, 'Our Big Happy Family'."

"Well then," says Nan, "maybe that's all that's really important."

"What do you mean?"

"I mean, why does it matter if you don't have all the pieces? We may not have a complete picture but that's OK, because we still know a lot about this family. We know they came to Deephaven, and had such a good experience that they had a puzzle made to remember and commemorate their vacation. And we have shared in their experience – by working together on their puzzle – just one big happy family – their family – and ours."

"You know," says John, leaning forward in his rocking chair, "It's all a big puzzle – isn't it."

"What?" I ask, knowing that John sometimes gets a little too philosophical.

"You see, no one really knows or understands what all the pieces are – or how they fit together, but that doesn't matter. We know what is important. It's family and relationships. That's what important. And you don't need all the pieces to understand that."

"I'll drink to that."

"Me too."

"Here here."

"To our big happy family," I say as we all sip our wine and munch on the bread and crackers. And we all realize ... I realize ... that John was right. That building relationships is what really matters. And as I put my arm around Nan who is sitting on the couch next to me, I realize that so much of my life has been wasted on things much less important – things that don't really matter. And so, I make a vow to myself – a vow to work harder on the things that I know are important. And I know I need to start now, because I may not have a lot of time left.

Nurturing relationships have always been difficult for me, and there at so many relationships that I know I need to work on. My relationship with my family and my relationship with Nan – to name two.

And also, my relationship with God. And even though I don't understand exactly who God is or what role He plays in this strange world of ours, I have begun to put some of the missing pieces together, and I am determined to forge ahead, and use the things I have learned to make a better life for me – and for my family.

And as we all sit there, and slurp our wine and munch on the crackers and bread, it's almost like we are in church – taking communion – the bread and the wine – and I feel as if there is a bond that is growing between us. A oneness. Something holding us together like never before. And I don't know whether or not God is playing a role in developing our relationships, and bringing this family together, but after this week, I am much more open to accepting things that I once thought were foolish and unscientific. And now I'm on a journey, and I'm not sure where I am going or what the final destination will be, but I have taken the first couple steps, and I know I am going in the right direction.

And as the dinner bell begins to ring, Mark says, “And I would like to make one more toast.” and he raises his glass and says, “To our big FAT happy family. Let's go to dinner.”

The Last Supper

As I walk into the dining room holding hands with Nan, I realize that our week together here at Deephaven is almost over. Tomorrow morning we will all eat breakfast, and then leave – going back to our old lives and our old routines. And I wonder, will anything really change? Will we revert back to our old ways, our old habits, when we return home? Has this week really made a difference, or is it just an aberration?

I don't know the answers to those questions.

I go up to the buffet and stack a plate high with meat and potatoes and set it down at our table. But before I sit down, I think that I should add a slightly more nutritious salad as my first course, and so I walk up to the salad bar in search of some lettuce, fresh vegetables and other sundry rabbit food – so that I can fool myself into thinking that I am eating a healthy meal. But then I see that they have homemade chicken noodle soup. One of my favorites. It smells so good. So, I abandon the salad idea and I use the ladle in the kettle to scoop up more soup than I will probably be able to finish, and then I begin walking back to our table holding the slightly overfilled bowl with both hands. But I don't get very far when a small boy runs by me and hits my leg as he passes. That causes me to stumble slightly, and some of the soup spills on the table to my left. Everyone at the table jumps up from their seats, but fortunately, the hot broth only falls on the tabletop, and not on to anyone's lap. I apologize to everyone and grab some napkins from the bar to help clean up the mess.

And that's when I see him.

There's a picture frame in the center of the table with a black and white picture of an old man with jeans and suspenders. And I recognize him immediately. I'm pretty sure he's the gentleman that helped me up the trail. The one that picked me up off the ground and held my arm as we walked to the top of Rattlesnake Mountain. At least, I think it's him.

"Who is that man in the picture?" I ask.

"Oh, that's our Uncle Bill. He's the reason our family is here this week."

"Do you know how I can get in touch with him? I need to talk to him – to thank him."

"Oh, that may be a little difficult to arrange. You see, Uncle Bill died last month."

"Oh no."

"Yeah, but it's OK. He had been very sick for a long time, and I'm sure he's much happier now."

"Oh ... well ... I'm sorry. I guess I must be mistaken. I must be thinking of someone else. It's just ... he looks so much like someone I met early this morning."

"Yeah, it must have been someone else."

"Yeah."

"Actually, the reason we are here is that Uncle Bill requested that his ashes be spread here in the mountains – because he loved this area so much. He spent much of his retirement just hiking up and down all the trails here, and he wanted his cremains to remain on Rattlesnake Mountain after he passed on."

"Yeah. This is a special place."

"Yes, it is."

And as I walk back to our table, I wonder, could it possibly have been Uncle Bill that helped me up the mountain? I don't know, but I do believe it's a possibility – a possibility I would never have entertained before this week. And as I take a seat, I realize that there might be a reason for everything that has happened here at Deephaven. I don't know exactly what that reason is, and I don't know where all this is leading. But I do know one thing.

I have a new direction in my life – a new faith in things I cannot see, a new hope for the future, and a new love for the people that matter to me the most.

And now abideth faith, hope and love, these three.
I Corinthians 13:13a

Made in the USA
Monee, IL
06 November 2026

68085976R00178